Bakelite Jewelry
good • better • best

Donna Wasserstrom

&

Leslie Piña

4880 Lower Valley Rd. Atglen, PA 19310

Dedicated to you,
the reader

Library of Congress Cataloging-in-Publication Data

Wasserstrom, Donna.
Bakelite jewelry: good, better, best / Donna Wasserstrom
& Leslie Piña.
p. cm.
Includes bibliographical references and index.
ISBN 0-7643-0122-5 (hardcover)
1. Plastic jewelry--Collectors and collecting--Catalogs.
I. Piña, Leslie A., 1947- . II. Title.
NK4890.P55W38   1997
745.594'2'09043075--dc21   97-94
CIP

Copyright © 1997 by Donna Wasserstrom & Leslie Piña

All rights reserved.
No part of this work may be reproduced or used in any form
or by any means--graphic, electronic, or mechanical,
including photocopying or information storage and retrieval systems
without written permission from the copyright holder.

ISBN: 0-7643-0122-5
Printed in Hong Kong

Published by Schiffer Publishing Ltd.
4880 Lower Valley Road
Atglen, PA 19310
Phone: (610) 593-1777; Fax: (610) 593-2002
E-mail: Schifferbk@aol.com
Please write for a free catalog.
This book may be purchased from the publisher.
Please include $2.95 for shipping.
Try your bookstore first.

We are interested in hearing from authors
with book ideas on related subjects.

Contents

Credits **3**
Any Questions? **4**
Jewelry:
    Bracelets **89**
    Pins **105**
    Necklaces, Clips, Etc. **146**
Objects **161**
Bibliography **175**

Credits

    Many people helped with the preparation of this book, and some are not even aware of it — those collectors and the curious who have asked questions over the years. We felt the questions that were asked most often deserved answers and we would like to thank these anonymous people as a group.

    In order to illustrate these questions and answers, we have used items from large and small collections and from dealers' inventories. The majority of pieces without caption credits are, or once were, part of the inventory of co-author Donna Wasserstrom. Some of the borrowed pieces are not credited in the caption, because the owners wished to remain anonymous. We would like to extend our gratitude to these anonymous lenders as well as to Cynthia Barta of Studio Moderne, Rita Eldridge, Judy Fitzpatrick, Shirley Friedland, Ken Jupp, Paula Ockner, Beverly Ratner, Susan Ranellone from Glitz 4 You, Bob and Veronica Romero, Shaboom's collection, Patricia Vreeland, Shelley Wilburn, and Sheila Wolf.

    Thanks of course to photography associate Ramón Piña, who also managed to keep the dust off the black velvet most of the time; again to Paula Ockner for diligently proofreading — any remaining errors are ours — to Rodney Wasserstrom for his support and exemplary display of patience, and to Douglas Congdon-Martin and the staff at Schiffer Publishing for being wonderful to work with and for the great job they do.

Any Questions?

1. Isn't Bakelite just plastic?

Yes, Bakelite is just plastic...and Rookwood is just pottery, Lalique is just glass, Stickley is just oak, Kensington is just aluminum, Icart is just paper....

The term "art plastic" was coined by industrial designers in the 1930s.

Habitat, carved in Poland in the 1930s by Stanislaw Kucharczyk.
All Bakelite, 7-1/4" x 5-1/2". $550

2. Who is this book for?

Dealers in Bakelite hear the same questions from different people. Answering them intelligently, or answering at all, is not always easy. Many dealers are not knowledgeable about Bakelite, and they label a variety of other plastics as Bakelite. For example, Victorian shoe hooks have handles that are sometimes labeled Bakelite, even though phenolic resins were discovered long after buttoned shoes were worn. Similarly, some Victorian handbag frames are also mistakenly called Bakelite. To help clear away some of the confusion, we have included some of the pieces that are often mistaken as Bakelite.

Many Bakelite buyers are hesitant, because they fear making a mistake. Not only can the material be hard to positively identify, valuation and pricing according to quality and rarity can be even more difficult. Whether buying or selling, reliance on someone else can be unsettling. Having the answers to important questions can be more helpful and more enjoyable than having deep pockets. Both buyers and sellers, collectors and dealers, can benefit from knowing how to identify the material and recognize quality.

Although some of the pieces pictured in these pages may be difficult to find, most of the objects are accessible. Although the yard sale sleepers are fewer and farther between, they are still out there waiting for the Bakelite connoisseur to spot them. Antique malls, shops, shows, and flea markets may have higher prices than one-time private sales, but the unidentified and underpriced little treasures turn up there as well.

Our intention is to answer some of the questions with words and with pictures. Knowledge is not the privilege of a few dealers or collectors — it is for anyone who seeks it. This book is for anyone with an interest in owning, evaluating, or just enjoying some of the wonderfully captivating little decorative objects made from Bakelite.

Wood and Bakelite necklace with brass chain and Bakelite pin, both c. 1944. Necklace: $700; pin: $500

3. How are prices established?

Prices for Bakelite, as well as other collectibles, are much higher today than they were in the recent past, and tomorrow, prices are sure to change further. Pricing in this book is only a guide based on an existing market, and since that market is always in a state of flux, so are the prices. Regional differences, trends, and fashion also affect prices, so market values may vary according to place as well as time.

We have tried to represent this varied pricing by listing one [United States] dollar amount that seems to represent both the retail value and an actual retail point of sale price. When items in a group are not listed separately in the caption, the price range, i.e. $300-500, means that one object is valued at $300, one at $500, and others in between. It does not mean that each item's value ranges from $300 to $500. With knowledge and luck, you may find similar items for less than these retail prices — sometimes far less. But all of the best items are relatively rare and command high prices.

It is impossible to please everyone — for some, heaven is too perfect. We hope to please, however, those collectors and dealers who understand the imperfection of any price guide, yet need some guidance. The purpose of a guide is to differentiate extremes more than it is to set prices. It can help you to avoid costly errors or to recognize a real find, because it is more important to know the difference between a $100 and a $1,000 item than to worry about whether to pay $75 or $100 for one. This is as good a place as any for the disclaimer, so: *neither the authors nor the publisher are responsible for any transaction outcomes based on the prices in this book.* We do believe, however, that one wise transaction resulting from good information will be worth the cost of this or any book.

"Good, better, best" is one way of illustrating differences in quality. This is not to suggest that an item is the best there is — only that it is the best one in the particular picture. By comparing similar items, we can learn to look more carefully and to see more clearly. Some carving is good, more carving is better, and lots of crisp detailed carving is best. Pins are usually more desirable than clips of the same subject. Sets are more valuable than the sum of their parts: it is good to have carved cherries, better to have more than one item with cherries, and best to have the entire set.

Do not put too much stock in numbers — either quantity for its own sake or in prices. Collect what you like. But by learning the standards for recognizing quality, you will begin to appreciate new things. That is how collections grow and change and how tastes develop. As long as your focus is on quality and enjoyment, rather than on numbers, your collection will have real value.

Above:
Hinged bracelets with medallions, c. 1940, evaluated on color and quantity of carving.
    black with olive carved portion and silver metal "button," back hinged — good. $350
    red with black carved "button," back hinged — better. $375
    apple juice with carved burgundy medallion, side hinged — best. $600

Next Page Top:
Different thicknesses. The thickness of the Bakelite affects the value on pieces with equal amounts of carving. A thin bracelet with good carving will always have greater value than a thicker plain bangle of the same width.

Next Page Bottom:
Carved cherries.
    earrings. $45
    pin. $250
    bracelet. $250
    necklace. $450
    The complete set is valued higher than the sum of its parts.

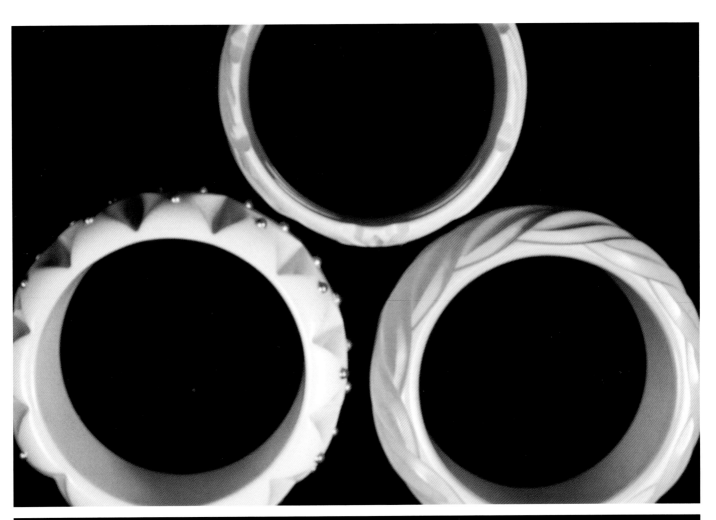

4. What is Bakelite?

Bakelite is a trade name for the synthetic resin, phenol-formaldehyde. It has become a generic term, however, for all phenolic resins — just as the brand name "Kleenex" refers to any facial tissue, or "Xerox" is used interchangeably with "photocopy." The term Bakelite will be used generically in this book.

Dr. Leo Hendrik Baekeland discovered the synthetic plastic while trying to produce synthetic shellac. When he combined carbolic acid and formaldehyde, the mixture would not pour out of the test tube. When neither heat nor any solvent would soften the new material, Baekeland realized that he had discovered something important — the first thermosetting plastic, a synthetic that would not melt once it had been heat set. Its initial use was in electrical insulation, and phenolic resins are still used today for that purpose.

In 1910 Baekeland founded the General Bakelite Company (later changed to the Bakelite Corporation), which became established in the United States, England, Germany, and Canada. Union Carbide acquired the Bakelite Corporation in 1939.

In 1925 the Catalin Corp. acquired the rights to produce a material called Herolith. When the Bakelite patent expired in 1927, companies raced to produce their own mixtures of phenolic resinous plastic. Catalin built its first U.S. plant in that year, and another larger facility in 1930. Then in 1932 Catalin purchased the cast resin patents of Dr. Fritz Pollak, and the name Catalin became synonymous for the brightly-colored phenolic resin.

Since "Bakelite" is the accepted generic term for any and all of the many phenolic resinous plastics, it will be used in that way in this book.

Detail of advertising ashtray from bottom of the piece, reads "Bakelite" and "TRADEMARK," c. 1928.

5. How can I tell if a piece of jewelry is Bakelite?

The easiest way is to touch the plastic with a red-hot needle. If it doesn't melt a little hole, it's Bakelite; if it does, you have a damaged piece of another type of plastic. Therefore, the test is not advisable until after you own the piece.

The next best ways to distinguish Bakelite from other plastics are by weight and the absence of seam or mold marks. Bakelite is relatively heavy. Purchase an inexpensive 1/2 or 3/4-inch bracelet to wear when shopping — this will aid in performing the weight test. Hold the Bakelite piece in one hand and the questionable piece in the other, then close your eyes and compare the weights. Next, look for seams or mold marks. Since Bakelite jewelry is made from cast blanks, there should be no signs of a mold. If you find a seam, it just isn't Bakelite.

Another method of identification is sound. Clink two Bakelite bracelets together and listen for a distinctive sound. Once you have become familiar with it, the sound test can be used on unknown pieces. You will also be able to tell if there is a crack in a bracelet — a broken piece will sound flat (like cracked or repaired pottery). Smell is another attribute of real Bakelite. Rub your thumb briskly along a smooth area. When it begins to feel warm, smell your thumb — it should smell like carbolic acid. Unfortunately, this method takes some practice and is not always reliable. You can also try to hold the piece under very warm water and then smell it, since wet Bakelite often has this distinctive odor. There is the possibility, however, that the formula used to make certain colors may prevent the smell test from working.

Patina is the natural surface change caused by exposure to the air and light over time. Like wood, metal, and most other materials, Bakelite acquires a patina. Put a dab of Simichrome polish on an old white sock and rub the inside or back of a piece of Bakelite. If the pink polish turns amber yellow on the sock, regardless of the actual color of the piece, it is Bakelite. However, not *all* Bakelite will create this reaction, so it is possible for the pink polish to stay pink, even if the piece really is Bakelite. Annoying? Yes. But anything worthwhile can't be too easy.

The best way to learn to identify Bakelite is to become familiar with all of the tests and to examine as much Bakelite as you can. Ask questions of dealers with large inventories. If you purchase an expensive piece, be sure to get a written guarantee, so that the piece can be returned if you discover that it isn't Bakelite. Study the pictures in books, such as this one, and become familiar with the colors, styles, and designs of Bakelite jewelry and other objects.

Bakelite identification checklist:
Hot pin test — will not penetrate Bakelite
Weight — relatively heavy
No mold or seam marks
Sound
Smell — carbolic acid odor if warm
Patina — Simichrome test
Visual — color and design

Early English Bakelite containers, c. 1928: humidor and box marked "Tea." $100 each

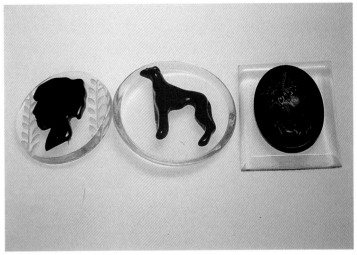

Three Bakelite with Lucite pins, c. 1940. *Courtesy Studio Moderne.* $40-55 each

Heavy, carved, wide bangles. Left: translucent red Bakelite, c. 1935; right: Lucite. *Courtesy Sheila Wolf.* Bakelite: $750

Stack of Lucite bangle bracelets with rhinestones, often confused with Bakelite. *Courtesy Judy Fitzpatrick.*

School necklace of celluloid, felt, and twine, c. 1938-42. $450

Plastic and rhinestone bracelet.

6. What is the difference between Catalin and Bakelite?

The Bakelite Corporation and the American Catalin Corporation were competitors in a dynamic market between World War I and World War II. Both companies used the same formula, for the raw material, but in a different form. Although the darker colors are usually considered to be Bakelite, and the bright colors Catalin, both companies produced both dark and bright colors, as well as pearlescent colors. American Catalin developed an exclusive process for inlaying metal in cast phenolic resin buttons. There were 110 trademark names for phenolic resins made by more than 100 companies; Marblette and Catalin were the most popular in the United States.

Note: New materials that simulate Bakelite are most often seen in black. Therefore, until you are very familiar with Bakelite, always test black pieces.

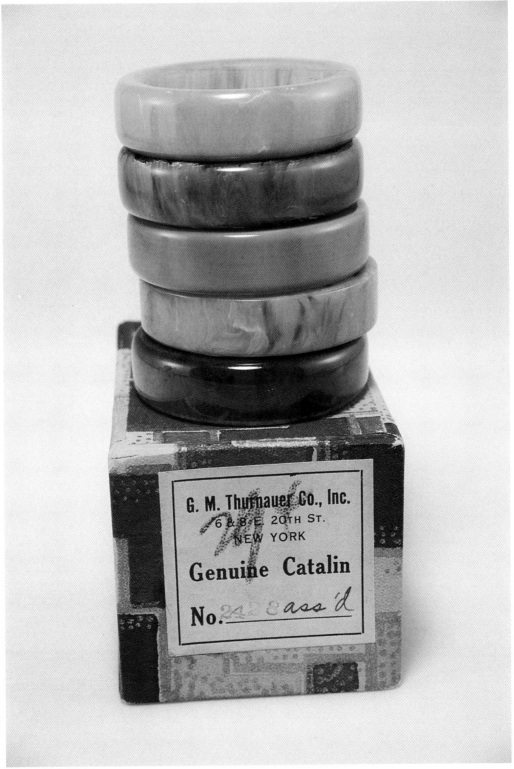

Napkin rings (one missing) in original box, c. 1930. Box is marked "Genuine Catalin." $80, with box

Octagonal lighter with rhinestones, in original box, c. 1935-40, also marked "Genuine Catalin." $85

These two bracelets, which closely resemble old Bakelite, are new and of a different material. They are cast and carved in the manner of old Bakelite, and when tapped they even sound like Bakelite. However, they feel slightly "greasier," and if you look closely at the carving, you will notice a white residue. There is no phenol odor when rubbed. When touched with a heated pin, it penetrated the material and gave off a very pungent smell, perhaps indicating a petroleum base. Similar pieces have been found in red and a "wannabe" apple juice color.

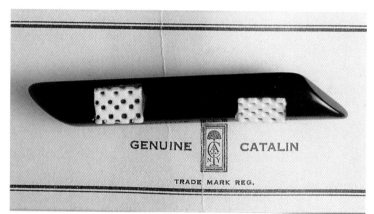

Catalin pin with inlay, on original card with Catalin trademark.

The top heart is contemporary and the bottom is Lucite.

7. What is the difference between Bakelite and other plastics?

Bakelite is a thermoset material, and once hardened, it is chemically and permanently changed. Other plastics, thermoplastics, can be resoftened and reformed. By 1935 the United States was the largest producer and user of synthetic resins. Plastics were once considered revolutionary, but by the early 1980s, world plastics production outpaced steel.

Plastics Time Line:

1868 **Celluloid** — the first plastic, made with nitric and sulfuric acids on cellulose, extremely flammable; the "Celluloid Era" was from the late nineteenth century through the early 1920s.

1897 **Galalith** — from the Greek *gala*, milk + *litho*, stone; made from protein derived from skim milk (casein) and hardened with formaldehyde.

1908 **Bakelite** — carbolic acid (phenol) and formaldehyde, the first entirely synthetic plastic, invented by and named after Dr. Leo Baekeland.

1919 **Casein** — hard horn-like plastic, first manufactured in American in 1919 (Galalith was earlier); when burned, smells like burnt milk or cheese.

1924 **Urea formaldehyde** — produced in the reaction of urea and formaldehyde.

1929 **Prystal** — clear phenolic resin.

1930 **Lucite** — acrylic trade name.

1930 **Vinyl** — thermoplastic derived from gas, obtained from the reaction of acetylene and hydrogen chloride.

1938 **Nylon** — thermoplastic resin obtained from organic compounds called polyamides.

Bakelite cat head with googly eyes on Lucite body pin, c. 1940.
Courtesy Judy Fitzpatrick. $850

Prystal pin with encased flower motif, on original card with trademark.

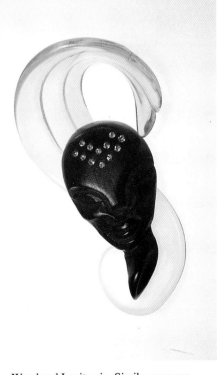

Wood and Lucite pin. Similar ones are often mistaken as Bakelite.
Courtesy Studio Moderne.

8. How can I distinguish Bakelite from celluloid?

Older celluloid was thinner and more brittle, besides being very flammable. Newer celluloid is somewhat less flammable. Celluloid can be molded into intricate designs, such as cameos. Since it is not a thermoset plastic, it can also be bent, twisted, and turned. Bakelite cannot. Most dresser sets are celluloid — look for rounded edges, designs requiring bending, and embossed patterns. Since celluloid is very flammable, the hot pin test is not advisable.

Detail of Bakelite pendant with celluloid cameo.

Celluloid clip. Bakelite cannot be twisted.

Celluloid and rhinestone pin.

Celluloid clip in Bakelite colors.

9. Is all Bakelite dark and all Catalin brightly colored?

No. Although Catalin was the first to introduce bright colors, both companies produced both dark and bright colors. There is usually no way to determine the manufacturer of an undocumented item, which most seem to be.

Note: A 1936 Fortune *magazine article referred to Catalin as "the gaudy brother of Bakelite." Catalin Corp. called its plastic "The Gem of Modern Industry."*

Pairs of bangles. Left: $250 each; right: $150 each.

Blue, green, red, and yellow necklace on celluloid chain with wooden spacer beads and matching elastic bracelet, c. 1940. $550 set

10. What colors are most often seen?

Shades of yellow, red, orange, green, brown, and black are the most common. Since air and light have created a patina on most Bakelite, the color may have changed. For example, many whites and light ivory tones have turned to amber or honey color.

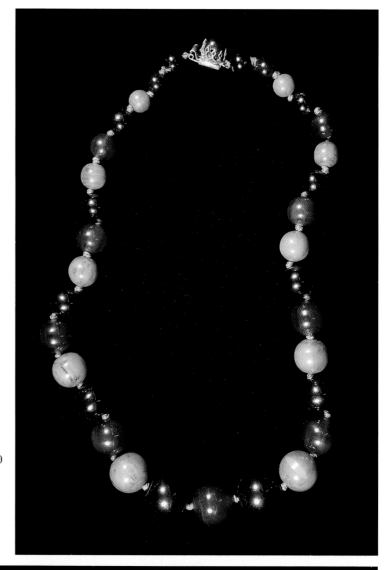

Right:
Multi-colored necklace, c. 1930. $150

Bottom:
Pairs of bangles. $150 pair

Facing Page:
Faceted bracelets. $75-200

Left: two clips; right: two pins, c. 1930s.
Courtesy Studio Moderne. Clips: $50 each; pins: $125 each

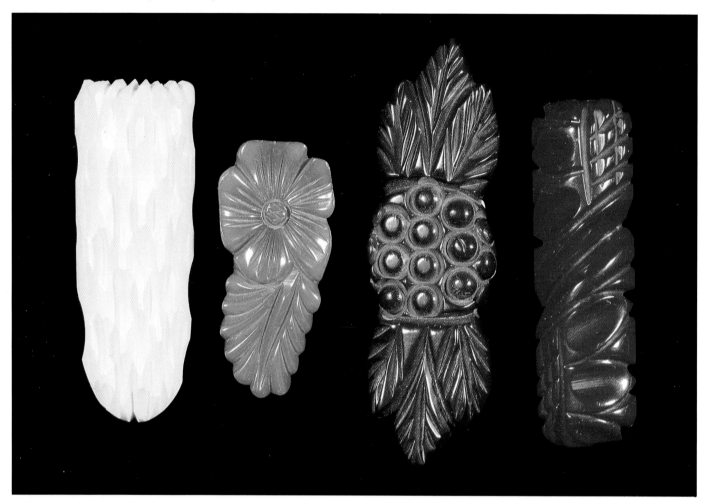

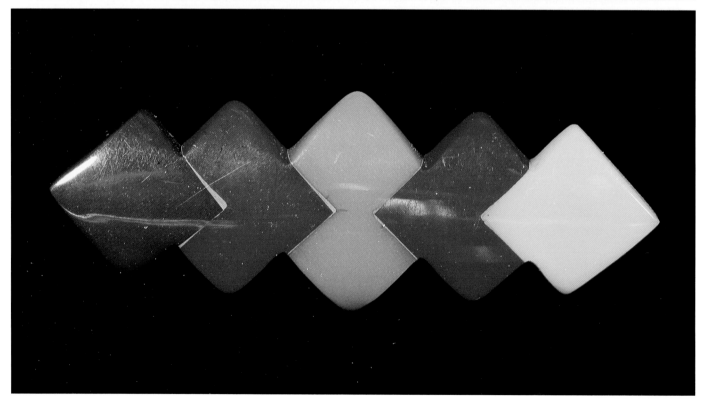

Five part pin in four colors, c. 1930. *Courtesy Studio Moderne.* $450

11. Why don't I see pink, turquoise, or white Bakelite?

These are just a few of the original colors that may have become patinized and are lying under the surface. Greens may be hiding shades of blue and turquoise; navy blue may be under what now looks like black; white may have turned to light yellow (cream corn); pink may be under orange; lavender may be under brown.

Refinished Art Deco perfume bottle of end-of-day Bakelite with Bakelite top, c. 1930. $75

New dots added to old bangles. $100-180

12. Why do many Bakelite colors sound like foods?

Names like "root beer, cream corn, butterscotch, apple juice, caramel" are used to describe colors of Bakelite. The associations have a tasty appeal, like the Bakelite objects.

Carved pin in root beer color. $100

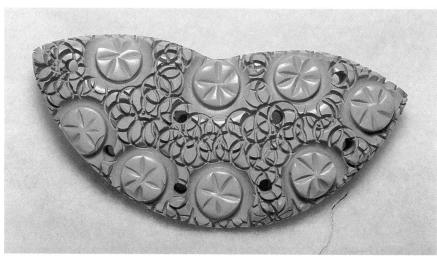

Heavily-carved butterscotch pin. $150

Square "Miami" pin in cream corn color with painted details, c. 1935-48. $500

13. What is resin wash?

When a piece of Bakelite has been soaked in a thinned lacquer solution, a translucent color appears on the surface. The most common resin color was red over butterscotch Bakelite, but other colors were also used.

Egyptian profile with brass details, c. 1930. $1,200

School pin, c. 1938. *Courtesy Beverly Ratner.* $650

Wood and resin wash Bakelite pin. $160

Contemporary turtle with "resin wash." $100

14. Why is there so much Bakelite and wood jewelry?

Production costs were lower when thinner sheets of Bakelite could be used, and combining materials was one way in which to accomplish this. It also created a different look, and added to the design possibilities.

In 1935 D. Lisner & Co. introduced wood and plastic laminated together without rivets or glue. They also imbedded wood in Prystal. Leo Glass made a higher priced wood and plastic jewelry for New York department stores, such as Saks and Bonwit Teller. After the Depression, some jewelry manufacturers used materials like aluminum, brass, high-glazed ceramics, and wood combined with Catalin.

Wood and Bakelite pins. $150-250

Wood and Bakelite pins.
    washerwoman with Bakelite pails — good. $90
    hat and Bakelite gun — better. $225
    gaucho with Bakelite hat — best.
    Courtesy Judy Fitzpatrick. $375

Wood and Bakelite bracelets.
    bottom: carved with design on wood to Bakelite portions — best. $250
    middle: bracelet with Bakelite corners — good. $100
    top: an animal holding a Bakelite disc in its mouth — better. $150

15. Was clear Bakelite ever produced?

Yes. For example, "Prystal," one variety of the clear material, was developed in the late 1920s. Today those clear pieces have turned a mellow amber color referred to as "apple juice." If it is clear today, it is not Bakelite. In addition, all translucent pieces should be carefully checked for internal fractures, because this type of Bakelite seems to have been more susceptible to heat damage than the opaque pieces.

Reverse carved apple juice bangles.
    top — better. $625
    middle — good. $550
    bottom — best. *Courtesy Judy Fitzpatrick.* $1,000

Apple juice bracelets, c. 1930s (except for bottom right with large black dots, which is not Bakelite and is contemporary). $300-650

Facing Page Top:
Celluloid chain with multi-colored Prystal drops, c. 1935. $200

Facing Page Bottom:
Apple juice pins. The best are those with reverse painting. $155-500

16. What is meant by reverse carved?

Some translucent pieces are carved on the back — the most common reverse carving is on apple juice. It was first done in 1937 by C.K. Castaing for trays sold by Georg Jensen, and the original designs included cacti and wheat. Reverse carved jewelry followed, and sometimes the carving was accented by painting.

Detail of reverse carved and painted bracelet

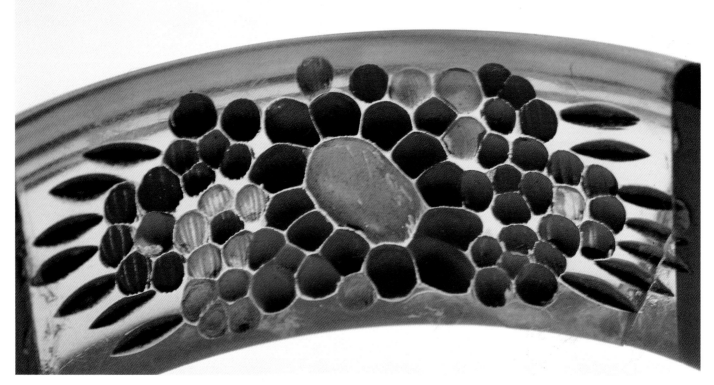

Detail of reverse carved and painted bracelet.

Detail of reverse carved and painted bracelet.

Detail of reverse carved and painted bracelet.

Detail of reverse carved and painted bracelet.

Detail of reverse carved and painted bracelet.

17. How is Bakelite laminated?

Pieces were bonded using a phenolic resin glue and then polished as one piece. The French were the first to experiment with laminating layers of Bakelite in 1935.

Stripes. Laminates consist of more than one color of Bakelite laminated to each other and sometimes carved and faceted after the lamination process. The more colors, the more valuable. $100-350

Two-color bracelets. These are also laminated. Some are made by casting one color over the other and then carved to reveal both colors. $100-250

Laminated pins. The largest is in two colors, the crest has three colors, and the peacock is multi-layered and made in France. $250 each; crest: $550

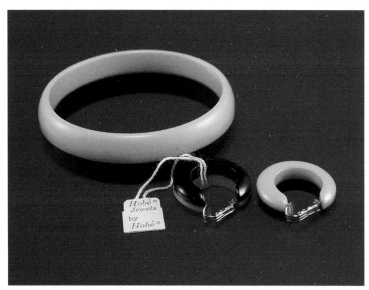

Two-color laminated bracelet, with matching earrings and original tag. The tag reads: "Jewels by Hobé Style V6B $2.25."
Courtesy Judy Fitzpatrick. $200 set

Carved and laminated bracelets (the second from the bottom is actually a dark navy, rather than black like the others). From top to bottom: $550, $350, $550, $150

Detail of carved, laminated bracelet.

Detail of carved, laminated bracelet.

Detail of carved, laminated bracelet.

18. How can American Bakelite jewelry be distinguished from European?

Since most Bakelite jewelry is unmarked, this can be difficult. Bakelite combined with metal, such as chrome or brass, can often be identified by the proportion of Bakelite to metal. European designs favored primarily metal with Bakelite accents; American designs used mostly Bakelite, with metal accents. Style can help to identify place of origin, since Europeans generally used more sophisticated streamlined designs, and Americans were preoccupied with novelty.

Jungle cat pin with finely detailed carving, European, c. 1930. $350

French necklace with Bakelite "perfume bottles," c. 1930. $475

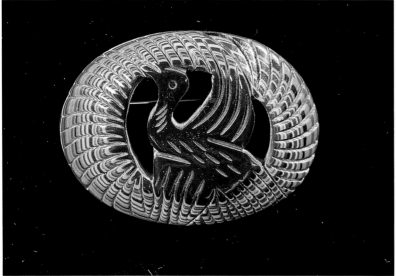

Laminated pin made in France. $250

Bakelite and brass cuff, made in France, marked, c. 1935. $150

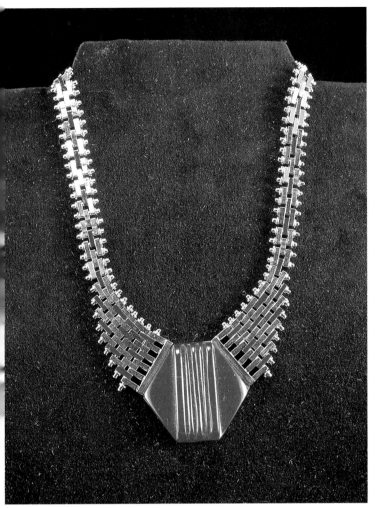

Above Left:
Outstanding machine-age enameled chrome and carved Bakelite necklace with articulated links, German, c. 1930. $1,200

Above Right:
Machine-age chrome and Bakelite necklace, German, c. 1930. $350

Left:
Galilith and chrome necklace, German, c. 1930. $75

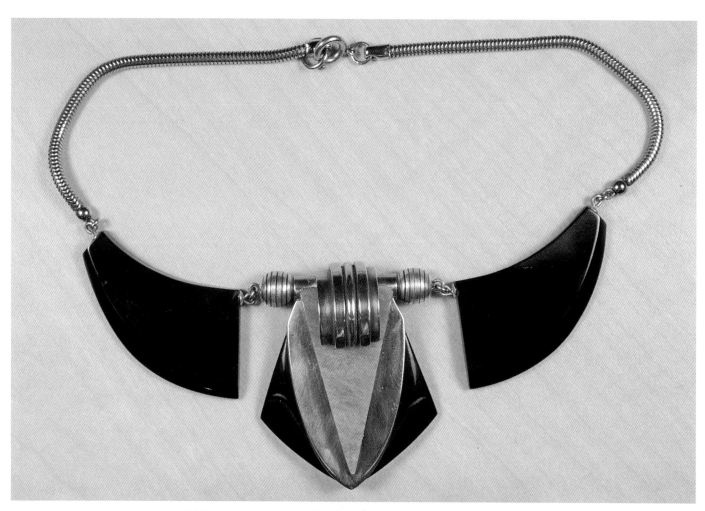

German Art Deco necklace of chrome with green and black Bakelite, c. 1930. *Courtesy Sheila Wolf.* $1,000

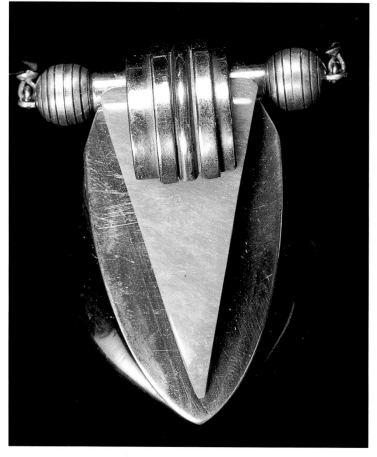

Detail of Art Deco necklace.

19. When did Pierrot first appear on jewelry?

The designer Elsa Schiaparelli popularized Pierrot in 1938, and some Bakelite bangles can be found. In 1975 the British costume jewelry designers Butler & Wilson made a line of pins and bracelets with the Pierrot theme. However, their pieces were Galalith rather than Bakelite.

Pierrot bracelets. Left, Bakelite, c. 1938; right, Galilith by Butler & Wilson, c. 1975. Left: $450; right: $250

20. How was Bakelite carved?

Pieces were carved on a lathe and finished by hand. Complex carving was often accomplished by more than one person in an assembly-line manner. Simple pieces could be polished by tumbling, like stones; detailed pieces were polished by hand using wet pumice followed by wax.

Carved Bakelite pendant on celluloid chain, c. 1930. $125

Previous Page:
Stack of carved bangles (top one is hinged), c. 1930s. From bottom to top: $450, $150, $250, $50, $350.

Above:
Stack of carved bangles, c. 1930-42.
Courtesy Studio Moderne. $100-150

Overleaf: Detail of carvings.

Detail of carving.

Detail of carving.

Detail of carving.

Detail of carving on bead.

Detail of bead.

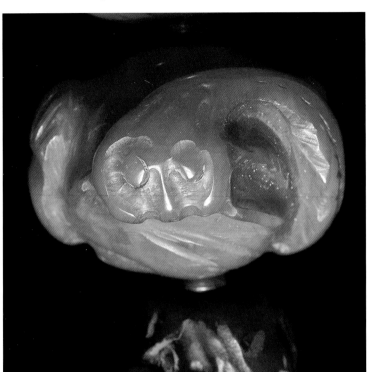
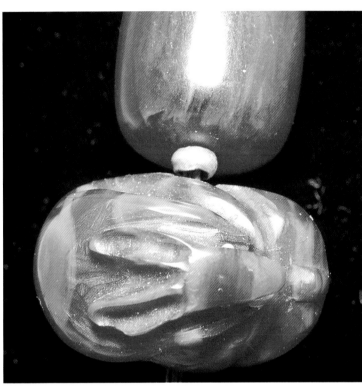

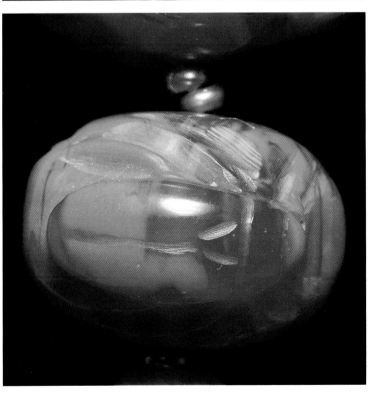
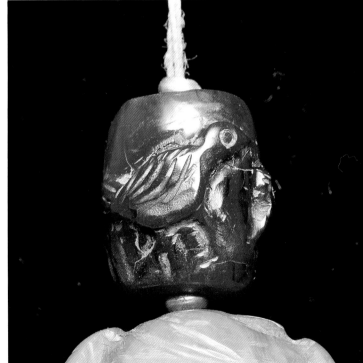

Detail of bead.

Detail of bead.

21. Is "end-of-day" Bakelite made like "end-of-day" glass?

Yes. At the end of the day, the leftover colors are poured and swirled in a mold, giving the resulting rod or tube a variegated appearance.

Carved perfume bottle pin, c. 1935. $225

End-of-day Bakelite blank, unfinished bracelet.

End-of-day green and brown Bakelite box, c. 1930, 3-7/8" diameter. $350

22. If white Bakelite turned yellowish, what is the white on my patriotic piece?

It is probably urea formaldehyde, sometimes known as "Beatl." Urea formaldehyde uses ammonia and carbon dioxide instead of carbolic acid. It was discovered in 1924 and substituted for some Bakelite when manufacturers realized that white Bakelite would change to yellow, and blue Bakelite would turn black.

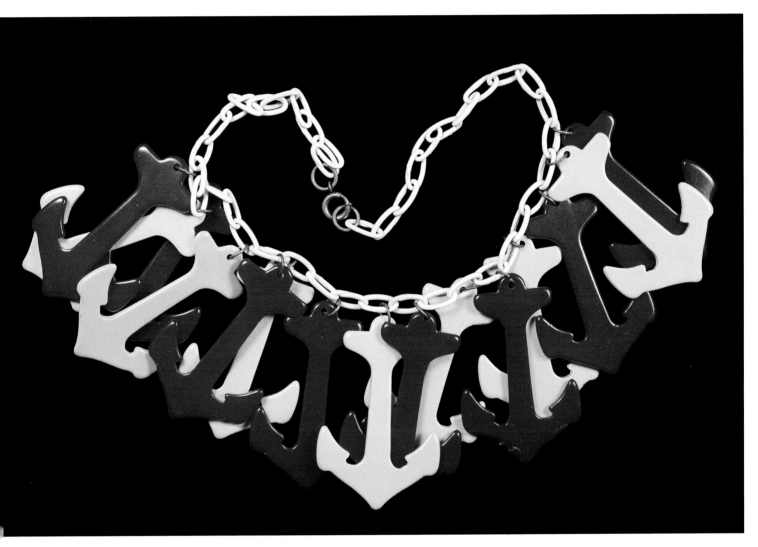

Anchor necklace in two colors, originally red and white, c. 1940. $575

Necklace, originally red and white, c. 1940s. $250

Pin with red and cream charms, c. 1935. *Courtesy Beverly Ratner.* $450

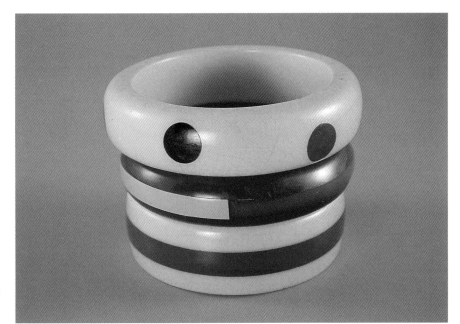

Three bracelets, originally red, white, and blue, c. 1930-42. *Courtesy Judy Fitzpatrick.* $250-450 each

Left: Bakelite laminated stripes with urea formaldehyde (white) chip. Right: pin of urea formaldehyde with Bakelite berries. *Courtesy Shelley Wilburn.*

Wood with urea formaldehyde. The colors are wrong for Bakelite. *Courtesy Shelley Wilburn.*

23. Can patriotic jewelry be dated?

Yes. It was made c. 1941 using American themes, such as insignia of each of the armed forces, Uncle Sam, eagles, flags, victory signs, and servicemen dressed in uniforms.

Cover of April 28, 1941 issue of *Life* magazine showing Bakelite pin called "MacArthur Heart." It was available in three sizes and butterscotch color.

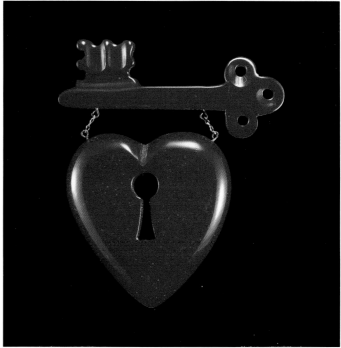

Large "MacArthur Heart" pin. $2,500. This pin was originally sold in the early 1940s for $1.00 on a red, white, and blue card which read, "He's the Key to Our Liberty and His Heart & Soul is With Us."

Patriotic pieces, early 1940s. The heart pin is made of Bakelite on top with urea formaldehyde white and blue layers.
 heart — good. $200 drums — better. $295 wings — best. $650

World War II patriotic pins:
 red spread wing eagle — good. $185
 painted V for victory — better. $350
 red, white, and blue shield with berries — best. $550

World War II patriotic pins. *Courtesy Judy Fitzpatrick.*
 heart $300
 heart from bar $600
 bar $250
 drum clip $350
 anchor $225
 majorette hat $150
 victory $450
 V from bar $400.

47

24. Why is the Scottie dog such a popular theme?

The Scottie became popular beginning in 1933 when Franklin Roosevelt was President. His Scottie, called Fala, was a national mascot. Asta, the wire-haired terrier from the *Thin Man* series also helped to popularize the dog theme.

Scottie on disc, c. 1933.
Courtesy Rita Eldridge. $400

Wood and Bakelite Scottie, c. 1935. $250

25. How is metal used to decorate Bakelite?

Bakelite can be plated. The piece is coated first with a varnish, followed by a conducting agent. It is then submerged in an acid bronze stop bath to prepare it for a 14k gold, silver, or copper surface. The actual plating is done by an electrolytic bath, and final polishing is done by hand.

Pin and two bracelets, excellent examples of plating over Bakelite.
Courtesy Beverly Ratner. Pin: $150; bracelets: $300 pair

Detail.

Two Bakelite bangles with silver deposit. $125 each

26. Is gold ever used with Bakelite?

Yes. Some apple juice Bakelite has particles of real gold or silver suspended inside. It was produced in 1936 and called "Star Dust."

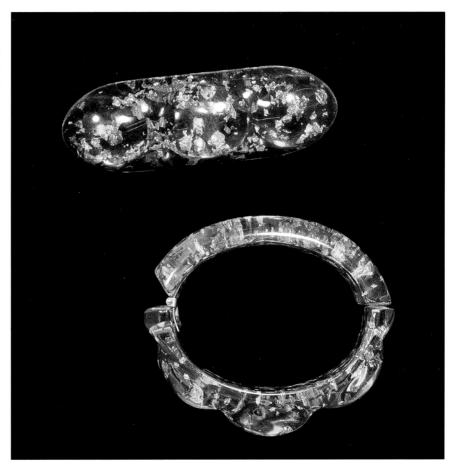

Apple juice bracelet and matching pin with flecks of gold leaf, c. 1938. $750 set

27. What are "bowtie" bracelets?

Bowties were made by inserting rods of one or more colors into the width of a bracelet — the resulting design looked like bowties.

28. What are "gumdrops?"

This is sometimes another name for bowties, but gumdrops usually refer to bracelets with randomly placed dots of various sizes. Gumdrops made in the 1960s and 1970s were made by injection molding.

Contemporary Italian cotton fabric with Bakelite bowtie and gumdrop bangles.

29. What are "6-dots?"

A 6-dot is a bracelet about 1/2 inch wide, with six evenly-spaced dots that wrap around the sides of the bracelet, and made in the 1930s.

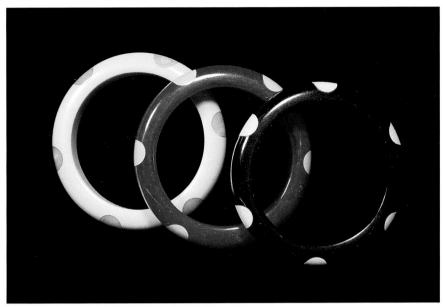

Three 6-dot bracelets, c. 1935. $350-450

30. What are Belle Kogan bracelets?

Belle Kogan designed for Blefeld & Goodfriend, who were jobbers for F.W. Woolworth's. She redesigned two-color polka dot bracelets by elongating the dots or making them oval shaped. One of these designs has only two very elongated dots, each being almost 1/2 the length of the bangle.

Stacks of dots, some inlaid, some injection molded, some applied, 1930s-1950s. Belle Kogan dots are in the right column, 4th from the top and bottom.

31. Why does the pin in the book have four charms, and mine only has three charms?

The basic pin was cut from a pattern, but charms were attached by different individuals, so variations occur. Number of charms can affect the value of an item. Figural charm bracelets should have at least five charms.

Resin wash over butterscotch Bakelite with brass tack and glass eye; matching six-charm bracelet, c. 1935. $195 each

Horse charm pin, with matching charm bracelet with brass chain and brass studs on horseshoe charms, c. 1935. $295 each

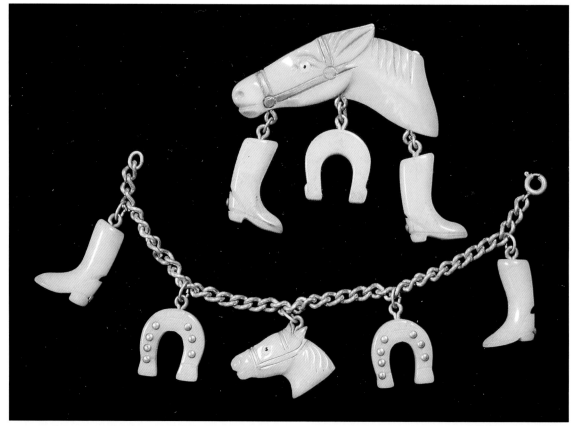

Banana pins, c. 1935.
  Right: all carved.
  Left: carved and painted and probably a knock-off by another manufacturer (also missing one banana).

It was not unusual for a manufacturer to produce two different sizes in a piece of jewelry. Small: $180; large: $275

Similar pins with different charms, c. 1930. $600-650 each

32. Who was Martha Sleeper?

Martha Sleeper was an actress who also designed novelty jewelry. The New York company, D. Lisner, manufactured her pieces between 1938 and 1942. These were clever and whimsical, usually combining Bakelite with other plastics or wood. For example, the bars of the birdcage on the "birdcage necklace" incorporate pieces of knitting needles. Her subjects included smoking items, school themes, cats, birds, and wishing wells; she also designed a line of buttons.

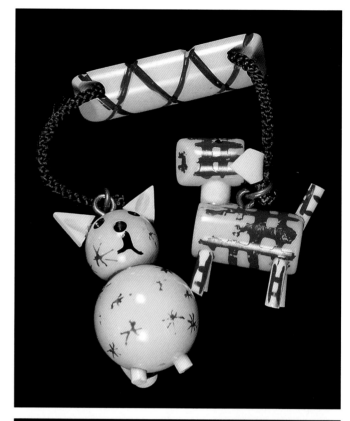

Calico cat and gingham dog pin by Martha Sleeper, c. 1938. *Courtesy Rita Eldridge.* $550

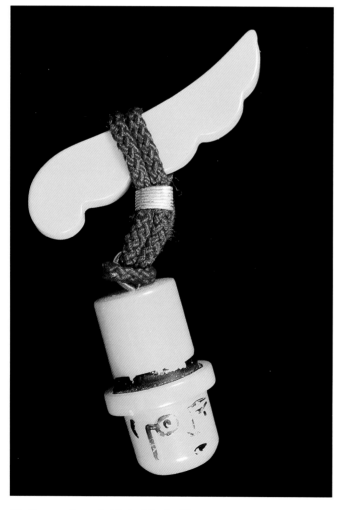

Mr. Peanut pin, probably by Martha Sleeper, c. 1938. *Courtesy Sheila Wolf.* $300

Martha Sleeper clip of drum with drumsticks, c. 1938. *Courtesy Judy Fitzpatrick.* $350

Martha Sleeper pin of birds on a branch, c. 1938. *Courtesy Judy Fitzpatrick.* $550

Martha Sleeper necklace and button, c. 1938-42. Button: $35; necklace: rare

33. Is Bakelite jewelry marked or signed?

Some pieces are marked with the country of origin, especially European pieces. The only designer known to have signed his pieces was the Frenchman Auguste Bonaz, in the 1920s. Although his designs are sought after and pricey, some are made of Galalith.

Note: While Josephine Baker gave gifts of specially designed Bakelite jewelry, she only wore real jewelry, which her admirers gave her.

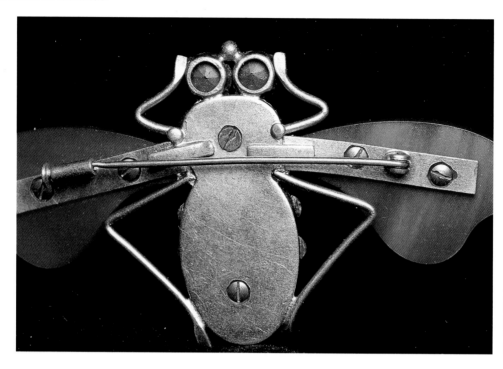

Back of pin. The back of the holder below the wing reads, "TCHECO SLOVAQUIE."

34. What are Schultz pieces?

Ron and Ester Schultz are among the most talented Americans to be making "new" Bakelite jewelry. They take old Bakelite from a bangle or a piece of sheet stock. Then they carve, cut, inlay, and laminate to create beautiful "new" pieces. These are proudly signed and represented as new. Rather than age, their merit lies in the design and artistry, and many Bakelite buffs assemble Schultz collections.

Large Scottie head, contemporary, signed "Schultz." *Courtesy Patricia Vreeland*

35. Why are some pieces so difficult to find?

Some of the complex pieces were never produced in large quantities. In fact, some designs were only made in a dozen or even fewer examples. These rarely appear in the marketplace, but remain in private collections, only changing hands among other serious collectors.

Pin of a frog musician with movable arm, c. 1940. *Courtesy Judy Fitzpatrick.* $2,000

Two-color Bakelite bracelet, c. 1950. *Courtesy Shaboom's.* $800

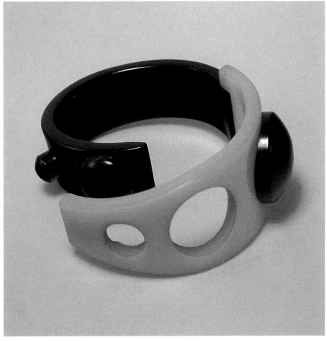

Two-color bracelet opened.

36. Why is so much Bakelite so expensive?

Rarity and supply and demand are the main forces driving up prices. There are only a finite number of surviving pieces, and new collectors continually enter the market. Generally, the most desirable and costly items are the ones that appreciate the most in value. Ordinary pieces are the least likely to gain value, and therefore should not be regarded as investments.

Fish bracelet and fish bowl pin, c. 1938. *Courtesy Beverly Ratner.* $1,500 each

Below: Detail of the pin.

This very rare carved necklace, c. 1930s, was made to look like amber. The carving is exceptional and probably done in Asia. Some of the beads are a translucent and opaque blend. The silver spacers are typical of jewelry from the Afghanistan area. A similar necklace in butterscotch is known to exist.

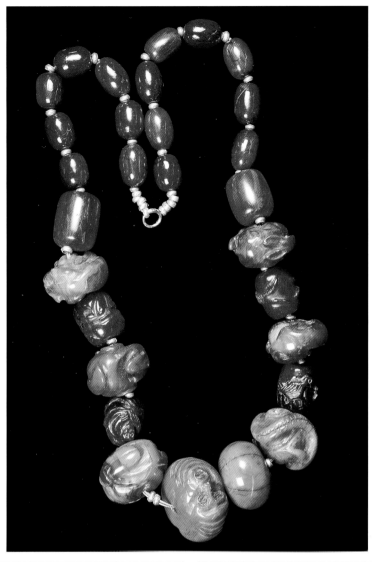

Detail of bead.

Detail of center bead.

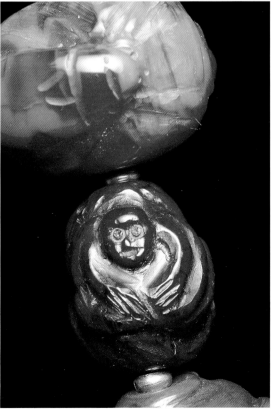

Philadelphia bracelet, c. 1935. In 1984, two of these bracelets were purchased for $250 each at the Art Deco show in Philadelphia, hence the name. Later it was also applied to matching pins and clips. *Courtesy Studio Moderne.* $2,500

Top view of Philadelphia bracelet.

37. What distinguishes a special and costly piece from a common and inexpensive piece?

Factors such as details, thickness of the item and its carving, color combination, novelty, subject matter, and attribution can affect price.

Note: During the Depression, Cartier designed a watch with a black Bakelite and yellow gold case. Westclox made a "handbag" watch designed by DeVaulchier & Blow, which was included in the 1934 "Machine Art" exhibit at the Museum of Modern Art in New York. In 1986 the Brooklyn Museum exhibit "Machine Age in America 1918-1941" also included Bakelite. In 1989 Sotheby's held a large sale of Bakelite items in Amsterdam.

Although inexpensive, these interesting bangles make good spacers, and should not be overlooked. $5-50

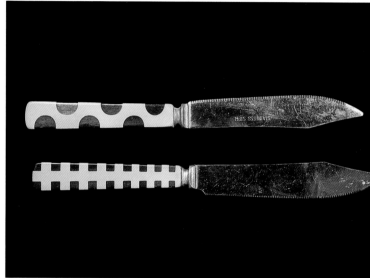
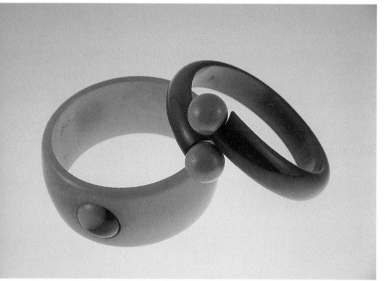

Top: Two-tone Bakelite cutlery handles, c. 1933. *Courtesy Studio Moderne.* $25 each

Bottom: Two unusual "dot" bracelets. Left: dots of three different colors, c. 1948. Right: balls on ends of bangle, c. 1930. Left: $150; right: $300

Top: Cutlery handles of two-color laminate Bakelite, c. 1935. *Courtesy Studio Moderne.* $75 each

Bottom: Left: laminated layer bracelet. Right: laminated two-color bracelet, c. 1935. Left: $800; right: $550

Left:
Fruit pins, c. 1935. *Bottom, courtesy Shelley Wilburn.*
$400-450 each

Below:
Mixed fruit on leaf pin, c. 1935-40. *Courtesy Judy Fitzpatrick.* $950

Bottom:
Crib toys, c. 1930s.
Better and best courtesy Bob & Veronica Romero.
    Left: good colors — good. $150
      Center: details of eyes — better. $275
      Right: unusual cowgirl with hat — best. $350

38. How can I tell old from new?

All Bakelite is old. So-called "new" Bakelite is actually old material fashioned into new pieces. Most Bakelite was produced before 1950, though some radio, television, and camera cases were made through the mid-1950s. Bakelite produced after World War II has less intense color than the earlier material.

Many artists are producing new pieces today by laminating, carving, or otherwise reworking old material. Since the older Bakelite has acquired a patina, any disturbance of the surface — carving or buffing in particular — will reveal the original vibrant color. On these newly worked pieces, the reds, yellows, and greens are just too clear and bright. Colors like white and turquoise are sure signs that the piece has been altered, because those colors have changed the most dramatically over time.

In addition, the surface should show some signs of wear and use. Tiny scratches, chips that have the same patina as the surrounding area, and a waxy mellowness are all indication that the piece is right. If the surface looks too new and shiny, it probably is. Occasionally, a perfect looking piece that had been carefully stored for decades will turn up, but most of the time, if it looks too good to be true, it is.

Note: In the 1930s, half of all jewelry sales were pieces that were mass produced phenolic resins. In winter of 1934, 30% were cast resins; by the spring of 1935, the percentage was 70%.

Above:
Contemporary horse pin, signed Schultz; contemporary cowboy pin, signed Schultz; and contemporary Indian Chief pin with detailed inlay. *Courtesy Judy Fitzpatrick.*

Left:
Brass signature, Diane Von Furstenberg, c. 1970s.

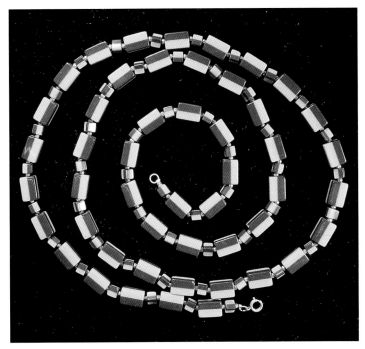

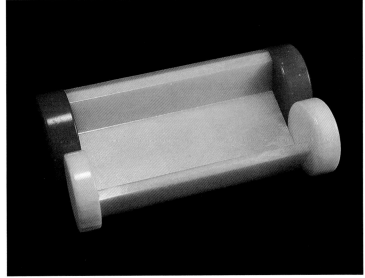

Top Left:
Necklace of injection molded Bakelite pieces, c. 1970. *Courtesy Shirley Friedland.* $80

Top Right:
Bakelite chip bracelet, c. 1950. *Courtesy Studio Moderne.* $65

Bottom Right:
Card holder with a base made from the short piece of a mah-jongg tray and the ends from checkers, c. 1950. $100

Bottom Left:
"Radio enhancer" made in 1950s in Cleveland, Ohio, of aluminum and Bakelite poker chips (although identified as Lucite), available in red, green, blue, and yellow. The Bakelite pieces were undoubtedly from a stock of uncirculated poker chips, as many items made after 1947 were made into unique items from older stock. $75 each

39. Where is the best place to buy Bakelite?

If you are a beginning collector, buy from someone who really understands the material. Find a dealer with a large and varied selection, and ask lots of questions. Feel the weight, listen to the sound, handle the piece. Be sure that your purchase is guaranteed to be Bakelite on the receipt. Later, when you have acquired some degree of connoisseurship, your knowledge can be more valuable than your checkbook. Garage sales, flea markets, and auctions are where the knowledgeable find bargains. But since these one-time events offer no guarantees, be prepared to learn from your mistakes as well. You should also be prepared to spend a great deal of time hunting and still, on occasion, return home empty handed. Garage sales aren't what they used to be. So you might also need to rely on specialist dealers after all. Even the sophisticated collector can benefit by cultivating a relationship with a reputable dealer.

Top: heart elastic bracelet. Left: heart hanging from a wooden bow pin. Right: heart wrapped in wooden bow. Bottom: heart with berries. *Courtesy Judy Fitzpatrick.* $125-250

Bird on wood branch — good. $300
Duck — better. $500
Crawfish — best. $650
Courtesy Judy Fitzpatrick.

40. What should I look for when shopping for Bakelite jewelry?

First, find something that catches your eye — that leaps out of the case for you. Then look carefully for possible problems: cracks or other signs of damage, deteriorated findings (clasps, clips, chains, pins, and hinges), glue or signs of repair, mold marks or seams, color changes on carved portions, colors that are too bright or new looking. If a pin has been glued on, it could be covering a spot that once held a button shank. If there are holes on the back, it may have been a clip or half a belt buckle in another life. If a piece has berries or charms and the strings have been redone, this is for protection and is not a problem; perfect strings are rare. If a bracelet has had the elastic replaced, the ends should be hidden. If a necklace has been restrung, the findings should still be original. If a celluloid chain looks cracked or dried, it could fall apart; if the chain looks too new, it just might be.

Look for signs of quality: rich mellow colors, indications of hand carving like tool marks, polishing in logical places and not in crevices. If you see signs of quality and none of the many possible signs of problems, then you've found a great piece. If you really like the piece and the price is right, it's probably your lucky day.

Translucent (Prystal) bracelet with sun or heat damage; note line the top.

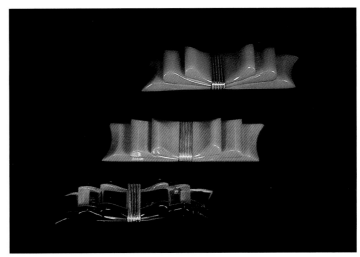

Three ribbon pens, c. 1935, one of laminated Bakelite. Note the fracture in apple juice section of bottom pin. Top two pins: $250

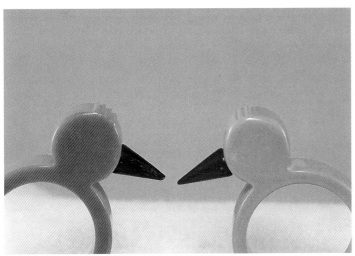

Details of two chicken napkin rings: the one on the left has a chipped beak, a common problem that adversely affects value.

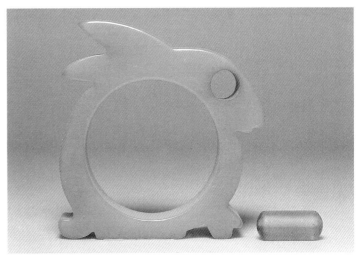

Small rod used as an eye for a napkin ring. If there is a hole in the napkin ring, this small rod has been lost.

41. Why do some expensive pins have "c" clasps instead of the better safety clasps?

The European clasp has a tubular cover that pulls away to release the pin. American versions are either stationary "c" clasps or the popular safety clasp, in which the pin is held in the center of a small wheel that locks closed. Since the vast majority of Bakelite jewelry was intended for a mass market, and much of it was sold in five-and-dime stores, the simple "c" clasps are commonly found. Though inexpensive in its day, the jewelry is judged by a different standard today.

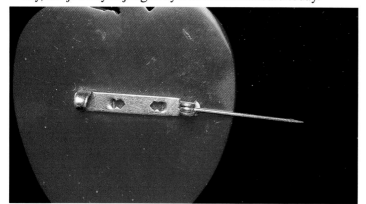

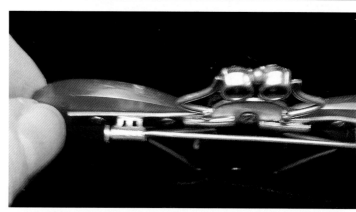

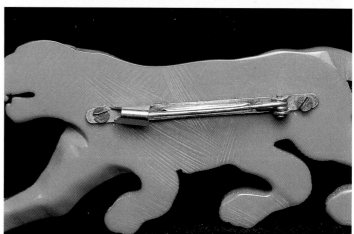

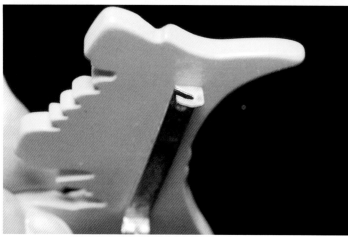

Top: Glued pinback. This may be a replacement, or one attached to an ex-button or part of a buckle or clip.

Middle: Pin back is inserted into the plastic base.

Bottom: C" clasp.

Top: Inset pinback. There should not be any glue where it is attached to the Bakelite.

Upper Middle: Riveted pinback.

Lower Middle: European pinback.

Bottom: European pinback on back of jungle cat pin.

42. Is any particular area of the country a better source for finding Bakelite?

Although most of the original manufacturing was in the Northeast United States, Bakelite jewelry can be found anywhere. Naturally, large cities with more shops and sources will have more selection, but prices may be higher due to overhead costs. Areas with larger populations of retired people may be better sources, but in general, Bakelite can be found across the country.

Jointed bell-hop pin, c. 1940. $750

Anchor pin with compass, c. 1940.
Courtesy Judy Fitzpatrick. $225

Pin made as inexpensive souvenir. $60

43. How can I determine when a piece was made?

Only some items can be attributed to a certain date. For example, Bakelite Union Jack jewelry was made in honor of the coronation of King Edward VIII in 1936; American patriotic themes were popular c. 1941; and the hinged bracelet first appeared in 1935. The following list will provide a general overview of popularity by decade:

1920s	1930s		1940s
geometrics and simple designs	cameos in bright colors	Hawaiian and tropical themes	Disney characters
bright dots	fruits, vegetables, flowers, leaves	multi-strand necklaces	fantasy characters
"Spanish" themes	pendants on long chains	clips	patriotic themes
cameos in dark colors	hinged bracelets	pets and other animals	rippling bows and ribbons
Egyptians, scarabs, cobras	bulky bangles	cowboys and Mexican themes	simple necklaces
Oriental themes	elastic bracelets	crib toy pins	narrow bracelets
Barbarics	reverse carving	Disney cartoon characters	felt or leather animal ears
red in 1929	ribbons, bow-knots	Martha Sleeper designs	"?" jewelry instead of dots
	lovebirds,	Belle Kogan dot bangles	"MacArthur Heart"
	arrows, sunbursts	generally bright colors	movable eyes, heads, and arms
	nailhead studs		
	berry-like beads		

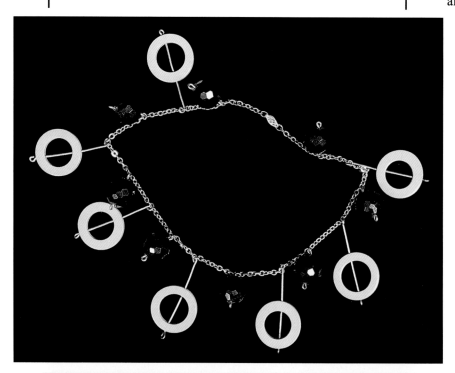

Art Deco Bakelite on metal chain, c. 1929. $175

Cameos of Bakelite and celluloid: the cameo is always celluloid because of the detailing, and the chain is celluloid. $120-175 pins, $150 pendants

Art Deco Bakelite and chrome necklace, also available in brass plate with Bakelite, and with matching earrings, c. 1929. $150

Medieval horse head, c. 1929.
Courtesy Rita Eldridge. $750

Heraldic lion pin, c. 1933. $750

Fortune teller, c.1938. $300

Group of Oriental inspired designs, c. late 1920s to 1930s. Two Chinese figures, piece marked Hobé, sword, and rickshaw. $175, $150, $300, $185 respectively

Assorted fruit pins, c. 1935-40.
    Queen Anne cherries — best. $650
    oranges — better. $450
    pears — better. $500
    bing cherries — good. $300

Thick butterscotch leaf pin, c. 1935. $385

Carrot bracelet. *Courtesy Beverly Ratner.* $350
Carrot pin. $600
Vegetable pins. $450 each

Crib toy pin, c. 1930. $1,000

Lucite and Bakelite birds pin, c. 1940. $250

Heavy black bangle, c. 1930. *Courtesy Judy Fitzpatrick.* $400

Deer with leather ears and carved dots, c. 1940 — good. $125
Bambi, c. 1935-42 — better. $195
Disney Bambi, c. 1935-42 — best. $250

Pins with berries, c. 1935-42. The green heart is contemporary, which can be determined by looking at the back of the pin. $250-450

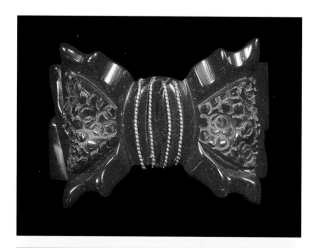

Thick red bow pin with brass detailing, c. 1930. $350

Ribbon bracelets, c. 1935-42. $325-375

Red nodder dog with swinging head, c. 1940. $850

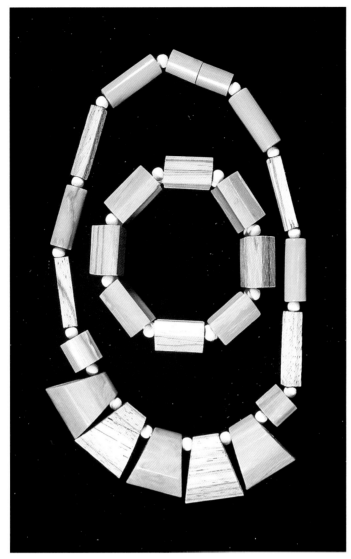

Bakelite and wood bracelet and matching elastic necklace with Bakelite screw clasp.
Courtesy Studio Moderne.
$100 set

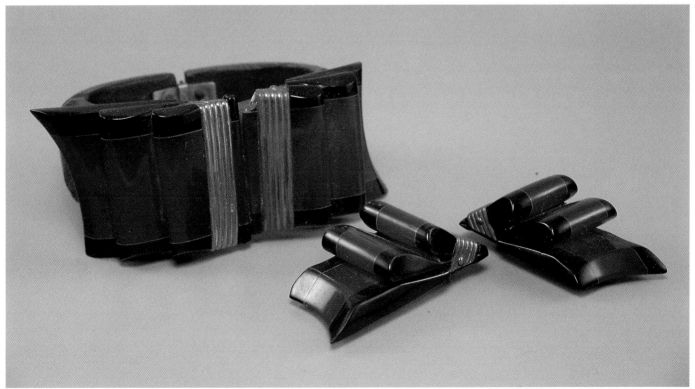

Two-color ribbon bracelet with hinged back and matching clips, c. 1935-42. Bracelet: $550; clips: $250

44. Why was Bakelite production ended, and when?

After World War II there was an economic resurgence. Women put away cheap plastic jewelry and other items that reminded them of the Depression and the war. Although Bakelite jewelry had been inexpensive, it was largely hand made, and this was impractical in the postwar world when production lines were being used to make nearly everything.

The manufacture of cast Bakelite blanks was discontinued shortly after the war. However, a large number of these blanks were warehoused and retrieved over the years. Today, some of the cast rods and tubes are still found.

Occasionally, Bakelite is still produced. In the 1970s, Diane Von Furstenberg manufactured a line of phenolic resin bracelets, but the colors lacked the depth and richness of the early ones. In the 1990s a few designers attempted to use new phenolic resins, but most recent accomplishments have been by using old material reworked into new items. Miriam Haskell issued a limited edition collection using old Bakelite, and Ron and Ester Schultz are known for their high quality new work.

Two watches with mechanical movements, c. 1940: green Endura, cream corn Royal Dynasty. $125 each

Butterscotch bracelets: the two in the middle are marked "dvf" for Diane Von Furstenberg and are from the 1970s; the two on the ends are from the 1940s. Left to right: $50, $65, $65, $150

45. Can I make my Bakelite jewelry shiny?

Yes. Bakelite can be polished using plastic polishes like Novus or with automotive store products like Turtle Wax and Simichrome. Use a soft cloth, such as flannel, and avoid over rubbing. This will prevent removal of the patina.

Carved red bangles. $150-350 each.

46. Will the value be affected if the patina is removed?

Yes. But if you prefer the original colors, personal preference can be more important than the monetary value. Eventually, a patina will form again anyway.

47. When were Bakelite containers for cosmetic and other household items used?

During the Depression, when most people could not afford to buy consumer goods, the field of industrial design developed in order to make items more attractive, desirable and therefore salable. Colorful plastic in the form of Bakelite containers was one solution for the under-consumption of certain goods. These plastics lost popularity after the war, and by mid-century, tougher acrylics took their place.

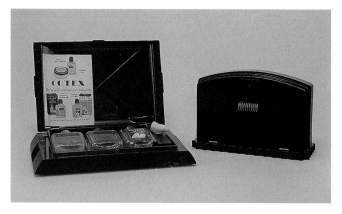

Travel containers. English soap with original foil wrapped soap; talc dispenser; and European talc shaker, c. 1930-35. $85 each

Top:
Two Cutex boxes, each with products and cotton. On the right box the front pulls down; the left box, which was never used, has a top that lifts. $45-75 each

Center:
Detail of box with lifting top.

Bottom:
Perfume bottles.
    De Corday with Bakelite top — good. $45
    Two of a set of three wood and Bakelite figurals — better. $30 each
    Set of two with glass dauber and replaced atomizer — best. $600 set

48. Where did amateur hobbyists get ideas for pieces?

In the 1940s, kits were sold in hobby stores that provided precut pieces or patterns to make them. Bakelite stock was also available, and magazines like *Popular Science Monthly* provided ideas for items that could be made at home.

Butterscotch with inlaid marquetry, probably a hobby piece of Bakelite and wood. $50

Home hobby kit which included blanks, an instruction booklet, a polishing tool, and patterns. Many pieces found today are truly one-of-a-kind, because they were made at home from materials purchased at hobby shops. Some of these home-made pieces are exceptional. Hobby kits were especially popular in the 1940s.

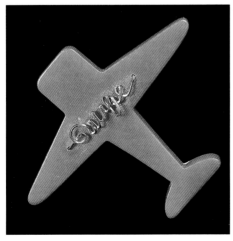

Novelty pin, probably a souvenir or hobby kit product. $85

Butterscotch box with cardboard and felt bottom, made from a hobby kit, 4-1/2 inch diameter. $125

49. Were any pieces originally painted?

Yes. Painting was a common means of filling in carvings, enhancing reverse carved pieces, and adding details to flatter pieces.

Detail of painted carving on bracelet.

Detail of pin showing painted reverse carving.

Painted hinged bracelet with blue painted over butterscotch Bakelite, c. 1940. $400

Rooster pin,
c. 1935. $350

Indian smoking a pipe,
pin of painted Bakelite,
c. 1936. $300

Two-color bracelets; the shaved area with color has been painted.
$100-300

50. When were Bakelite game pieces popular?

In the 1930s, Bakelite pieces for games like chess, checkers, mah-jongg, and poker were in vogue.

Matching necklace and bracelet of metal with Bakelite dice, c. 1930, and Bakelite earrings. *Courtesy Studio Moderne.* $125 necklace/bracelet set; $65 earrings

Backgammon doubling cube, 3-1/2 x 1-1/2 inches. $125

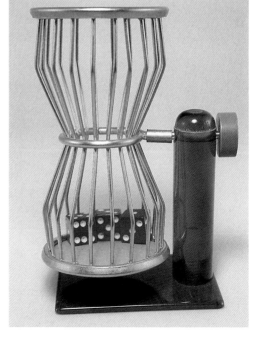

Chuck-a-chuck game piece and Bakelite dice. $225

51. Are there really "warehouse finds?"

Yes. Sometimes old blanks or parts are found — these are good. Other finds may include older material assembled into jewelry at a later date— these are also good, possibly better. Occasionally, uncirculated jewelry inventory from the period is found — these are the best.

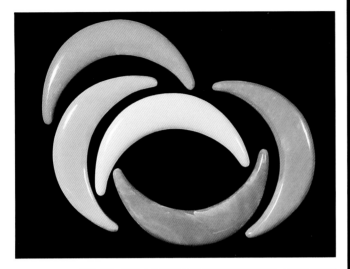

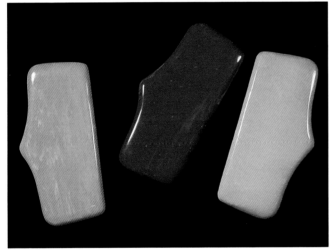

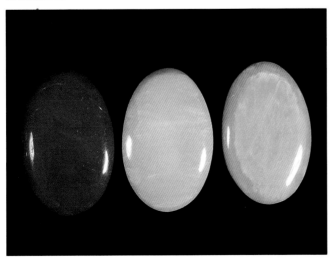

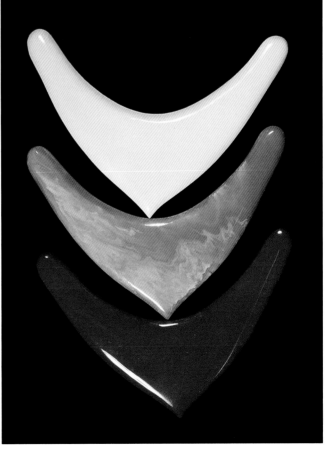

Bakelite blanks ready to be made into jewelry.
Courtesy Glitz 4 You.

Bakelite blanks ready to be made into jewelry. *Courtesy Glitz 4 You.*

52. Are plastic radio cases really Bakelite?

Yes. The first radio with a phenolic resin case was most likely a Silvertone sold by Sears & Roebuck. Bakelite cases enabled radios to become more affordable: dark colors often sold for $10 or less, and brighter colors for $15 or more. The FADA Streamliner radios of the 1940s were nicknamed "bullets." These streamlined designs often resembled automobile grills, because the industrial designers of the cars and the radios were often the same people. Later in the 1950s, light-weight and brittle plastics replaced Bakelite or radios, which are also highly collectible today.

Fada Model 1000 Bullet, Catalin radio, 1945 (a postwar version of the classic 1940 Model 115). *Courtesy Ken Jupp.* $700

Zenith brown Bakelite radio.
Courtesy Ken Jupp. $100

Zenith brown Bakelite radio.
Courtesy Ken Jupp. $100

Zenith brown Bakelite radio.
Courtesy Ken Jupp. $100

Stromberg-Carlson brown Bakelite radio.
Courtesy Ken Jupp. $100

Jewelry

Bracelets

We would like to introduce this section with a little story. An older woman was looking in my [Donna] showcase and inquired about the price of an especially nice Bakelite bracelet. When I told her it was $750, she told me that she had received a similar piece as a high school graduation gift. Teasingly, I said, "Don't you wish that you still had it?"

"I do," she replied. "I have all my Bakelite in shoe boxes under my bed."

I asked her as calmly as I could, "Would you like to sell any of it?"

"No," she answered, then hesitated and added, "...but I am going to change my will."

Of course, not all Bakelite bracelets or other jewelry items have such high monetary value. But many do. The examples that follow will help to illustrate the good, the better, and the best — there are no "bad" or "ugly" — and attempt to answer the "why."

Top:
Hinged bracelets:
    wood and green Bakelite
    — good. $550
    apple juice and maroon Bakelite
    — better. $650
    four colors of Bakelite
    — best. $1,000

Center:
Hinged bracelets:
    left: nice carving
    — good. $295
    right: ribbon bracelet with separate "ribbon" portion laminated to the bracelet
    — better. $400
    center: pear on wood with excellent carving of Bakelite and wood
    — best. $550

Hinged bracelet of "root beer" Bakelite, c. 1938. *Courtesy Judy Fitzpatrick.* $400

Above: Three well-carved bracelets, two with back hinges and bangle (center) deeply carved all around. *Bangle courtesy Beverly Ratner.* Left to right: $275, $400, $400

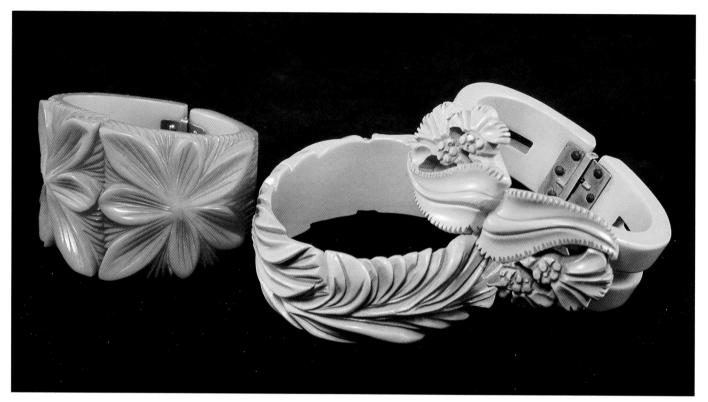

Three hinged bracelets, c. 1935-40, in which value is based on amount of carving and unusual design. Scottie: $750; others: $400. *Courtesy Judy Fitzpatrick.*

Two hinged bracelets, c. 1935-40.
right: Bakelite pears with wood leaves — good. $550
left: Bakelite gourd with Bakelite leaves — better. $700

Three hinged bracelets, all of exceptional design.
- left: Bakelite turtle on wood leaf — good. $400
- right: heavily carved top — better. $650
- middle: strawberries — best. $950

Detail.

Detail.

91

Elastic bracelets:
  right: black buckle, c. 1939 — good. $175
  left: multi-colored, c. 1935 — better. $250

Bakelite hinged bracelet resembling wood, with painted flowers, c. 1938. *Courtesy Judy Fitzpatrick.* $950

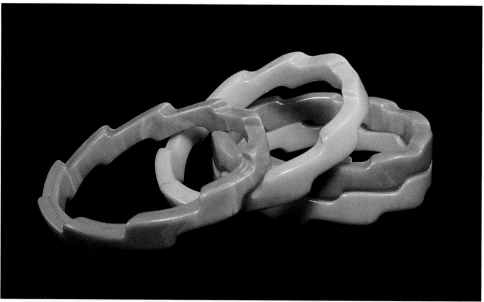

Set of four zig-zag bracelets, c. 1930. $500 set

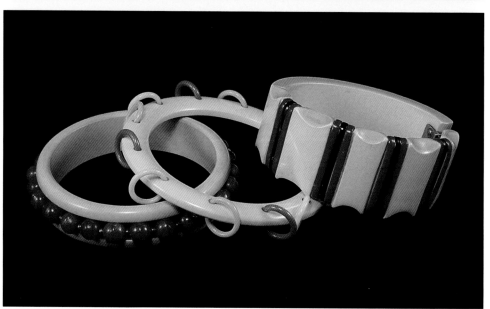

Bangles with unusual embellishments. $150-250

Green and butterscotch marbled Bakelite back-hinged bracelet, c. 1938.
Ringed bangle of four colors, c. 1935. Left: $250; right: $750

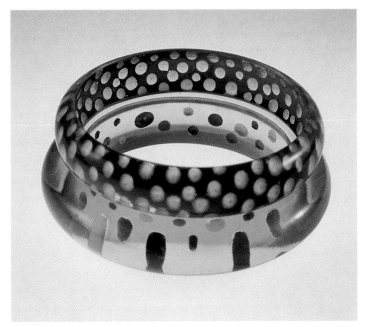

Reverse carved apple juice bangles with dots,
c. 1938. *Courtesy Beverly Ratner.*
    top — good. $250
    bottom — better. $400

Raspberry color carved bangle.
Courtesy Beverly Ratner. $400

Stack of reverse carved apple juice bangles, c. 1937-42. *Courtesy Judy Fitzpatrick.* $600 each; bottom $1,400

Two reverse carved bracelets.
top: dyed bangle — good. $600
bottom: hinged with carving on top and reverse painting — better. $700

Assortment of bracelets.
Courtesy Judy Fitzpatrick.
  dots. $500-700 each
  laminated. $300
  zig-zag. $500
  Bakelite and chrome. $250

Multi-color bracelets, c. 1930.
Courtesy Judy Fitzpatrick.
  bottom: two-color laminate
    — good. $250
  middle: two-color zig-zag
    — best. $400
  top: three-color laminate
    bracelet — better. $350

Similar group of green and cream bracelets. *Courtesy Judy Fitzpatrick.*

Various bracelets: 6-dot, laminate, and carved, c. 1930. *Courtesy Judy Fitzpatrick.* $400 each

Stack of bracelets, c. 1930, except third from the top, c. 1970s. *Courtesy Judy Fitzpatrick.* Bottom to top $200, $300, $550, $550

Stacks of bracelets with applied decoration.
    rhinestone — good. $50-150
    carved and painted — better. $50-75
    medal adornment — best. $75-300

Stacks of carved bracelets.

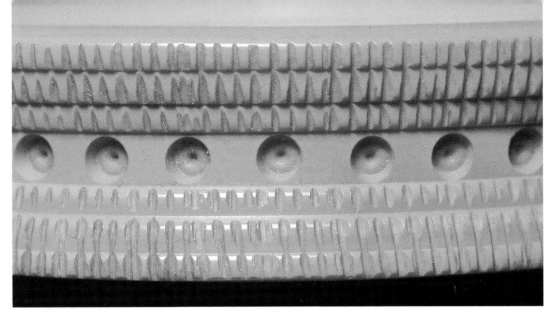

Detail.

Detail.

Detail.

Bracelets with heavier carving, c. 1935. $200-350

Detail.

Three carved bracelets, c. 1930.
Courtesy Judy Fitzpatrick.
bottom — good. $300
middle — better. $400
top — best. $600

Dots. The top and bottom bracelets are contemporary and not Bakelite. The large green dot is contemporary Bakelite; the yellow is too bright and has no patina.

Next Page:
Stack of bracelets. The bottom three are new. *Courtesy Studio Moderne.* The black and cream 6-dot: $450; others: $100

Cuff of translucent Prystal, c. 1935. $250

Bakelite and wood hinged bracelet, c. 1950. $100

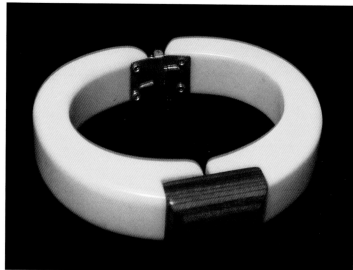

Bracelets of unusual colors.
    watermelon color Prystal, c. 1960 — good. $100
    cream and blue cuff, c. 1940 — better. $185

Bracelets, c, 1935-40. *Courtesy Studio Moderne.*
    bottom: plain — good. $35
    middle: cuff with rhinestones — better. $65
    top: back hinge with brass — best. $250

Pins

Bakelite material does not come shaded, so the bottom example is another material.
- middle: Bakelite and wood carved together — good. $175
- top: resin wash over butterscotch Bakelite — better. $275

Fish. Resin wash and apple juice are both desirable, but this apple juice example is large and also has silver paint on the reverse side.
- top: wood and Bakelite fish — good. $175
- bottom: resin wash sailfish — better. $275
- middle: apple juice fish — best. $300

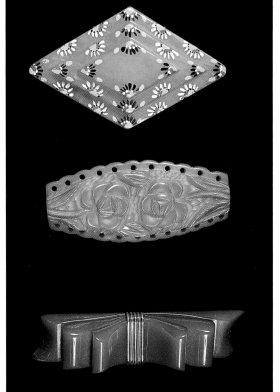

Pins.
- two-tone flower — good. $150
- large red carved pin — better. $285
- large black thistle with stamens — best. $400

Pins.
- painted layered Bakelite — good. $100
- well-carved butterscotch — better. $175
- ribbon pin — best. $250

Banana pins, c. 1935.
    single banana— good. $400
    cluster on leaf — better. $650
    three hanging from leaves — best. $850

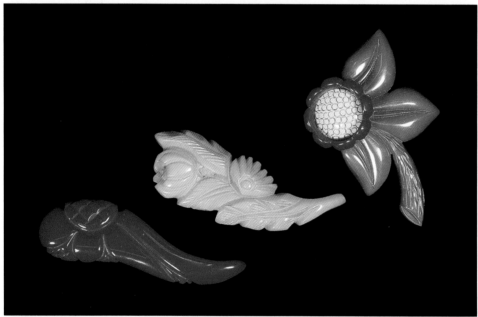

Flower pins.
    red carved — good. $125
    more carving — better. $175
    multi-colored — best. $350

Strawberry pins, c. 1935-40.
    strawberries on wood — good. $250
    apricot colored thick Bakelite — better. $300
    single large fruit — best. $500

Hitch-hiker pin, c. 1940. *Courtesy Judy Fitzpatrick.* $500

Jointed English Bobbie pin, c. 1930s. $650

Jointed figure pins, c. 1930s. *Courtesy Judy Fitzpatrick.*
right: sailor with non-Bakelite head and hat — good. $600
left: scarecrow — better. $900

Jointed figure pins, c. 1930-44.

Top:
Jointed scarecrow pin with yarn hair, c. 1935. $850

Center:
Jointed cowboy pin, c. 1940. $700

Bottom:
Jointed toy soldier pin, c. 1940. $650

Jointed clown pin, c. 1935. $395

Jointed Dutch girl pin with yarn braids and leather feet, c. 1940. $850

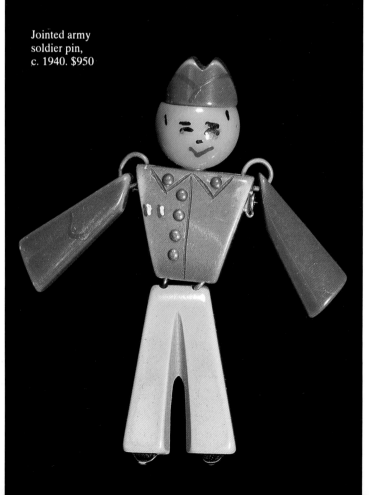

Jointed army soldier pin, c. 1940. $950

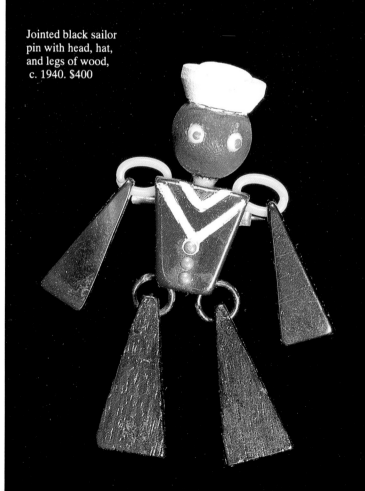

Jointed black sailor pin with head, hat, and legs of wood, c. 1940. $400

Animal pins, c. 1930s. The fox is a nodder with a movable head.

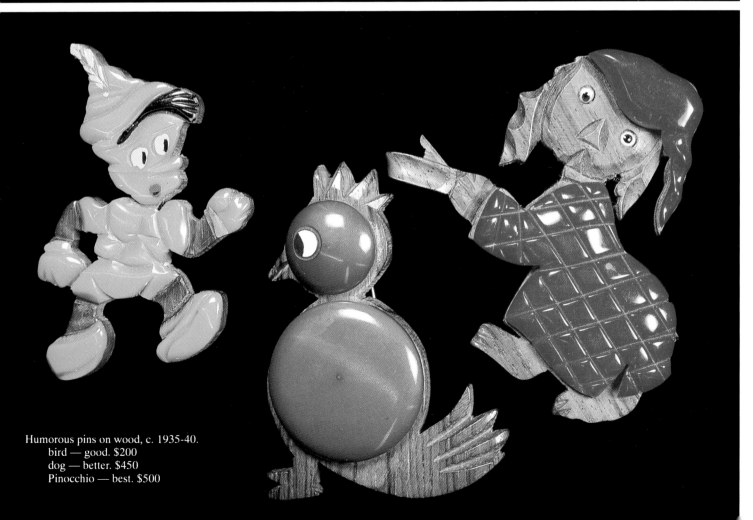

Humorous pins on wood, c. 1935-40.
    bird — good. $200
    dog — better. $450
    Pinocchio — best. $500

Pins with movable parts.
   fox with leather ears, c. 1940 — good. $450
   red frog, c. 1980 — better. $600
   marching soldier, c. 1940 — best. $1,400

Animal pins, c. 1930s.
   fish — good. $250
   horse — better. $395
   fox (with crack) — better. $395 (if perfect)
   heraldic lion — best. $800

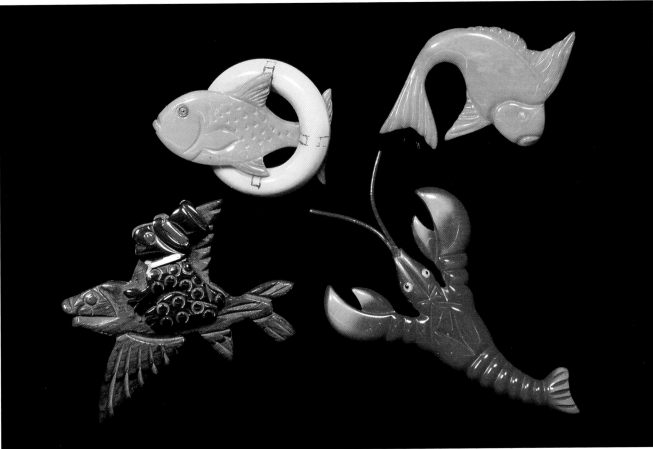

Top:
Pins, c. 1935, rated according to degree of uniqueness.
    horse head — good. $300
    gazelle — better. $600
    giraffe — best. $800

Bottom:
Sea creature pins, c. 1935-40.
    fish — good. $150
    frog on flying wood fish — better. $450
    leaping fish, carved to fit into life buoy — best. $950
    large resin wash lobster — best. $1,000

Red Bakelite alligator on wood pin, c. 1935. $125

Wood gazelle on Bakelite pin. $200

Center Left: Monkey with Bakelite hat and felt ears, c. 1940. $150

Center Right: Bakelite on wood pin. $400

Bottom Left: Prystal on wood dragon pin. $150

Bottom Right: Art Deco horse head pin, c. 1930, in marbleized Bakelite (which has a tendency to crack). $300

Below:
Polar bear pin with brass collar and chain, c. 1934, popularized because of Admiral Byrd's expedition to the South Pole. $800

Right:
Elephant pin, c. 1930. $450

Bottom:
Pins, c. 1935-40. *Courtesy Rita Eldridge.*
    Bakelit and wood rabbit — good. $375
    Disney's Thumper — better. $400
    Bakelite rabbits behind wood log — best. $450

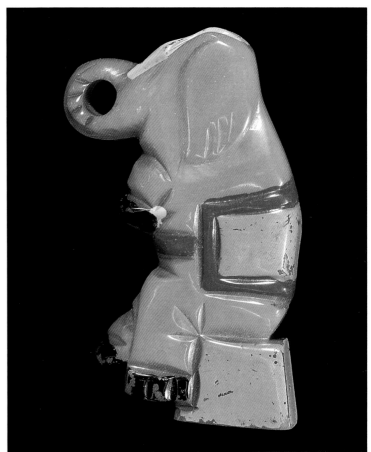

Merry-go-round animal pins, c. 1935. $350 each

Previous Page:
Horse and racing pins,
c. 1930-35:
    rocking horse $750
    Art Deco head $300
    race horse $850
    greyhounds $850
    boot with disc $350
    googly eye $850
    green head $325
    large butterscotch $700
    head with chain $375
    running on wood $195

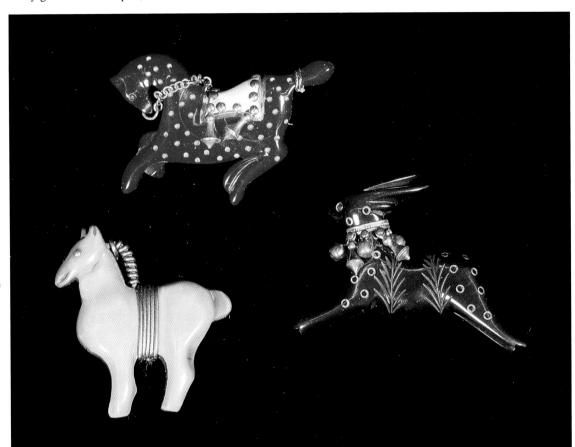

Horse head pins, c. 1935. $275-400

Dog pins $100 each, with leather ears $275 each.

Dog pins, c. 1930s. Scotties were popular because of President Roosevelt's dog, Fala. $100-150

Above:
Pins, c. 1935: *Courtesy Judy Fitzpatrick.*
    key and lock — good. $250
    arrow with hearts and beads — better. $400

Top Left:
Black lock and key pin, c. 1935-40. *Courtesy Sheila Wolf.* $250.
Black Scottie head pin, c. 1930. $125

Large heart with arrow pin, c. 1940. $900

Similar pins, but the bottom example has redone strings. Although original strings are preferable, if strings are replaced properly, it should not affect the value. $200-285

Bar pins with hearts, c. 1935-40.
    red hearts and butterscotch balls — good. $350
    Unusual black hearts — better. $600
    Mexican theme with jugs and sombrero — best. $750

Peacock pin, catalog 1935. $650

Owl pin, c. 1935-40. *Courtesy Judy Fitzpatrick.* $350

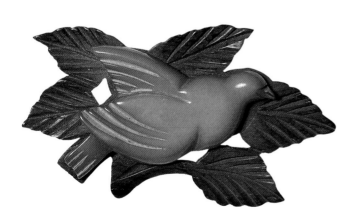

Bakelite bird in wooden branch pin. $150

Bakelite woodpecker on wood pin. $150

Bird pins, c. 1935-40.
    flamingo on leather — good. $150
    bird on wood branch — good. $150
    matching pair of birds on branch — better. $225

Birds on a branch, c. 1935. *Courtesy Judy Fitzpatrick.* $250

Insect pins, c. 1928-32
dragonfly — good. $125
Czechoslovakian with wood and rhinestone — better. $225
cicada — best. $500

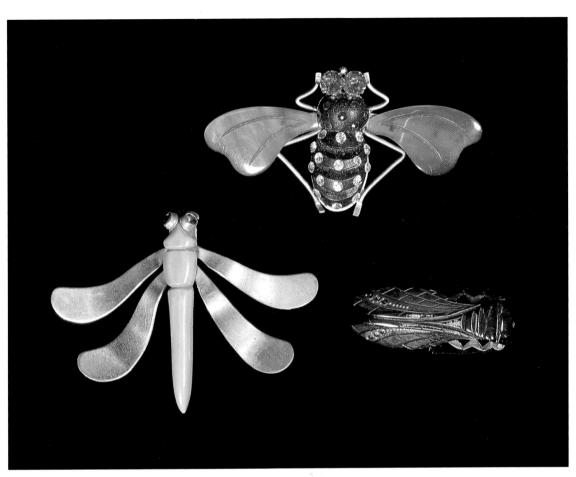

Spider pin with Bakelite body and brass, c. 1937. *Courtesy Studio Moderne.* $85

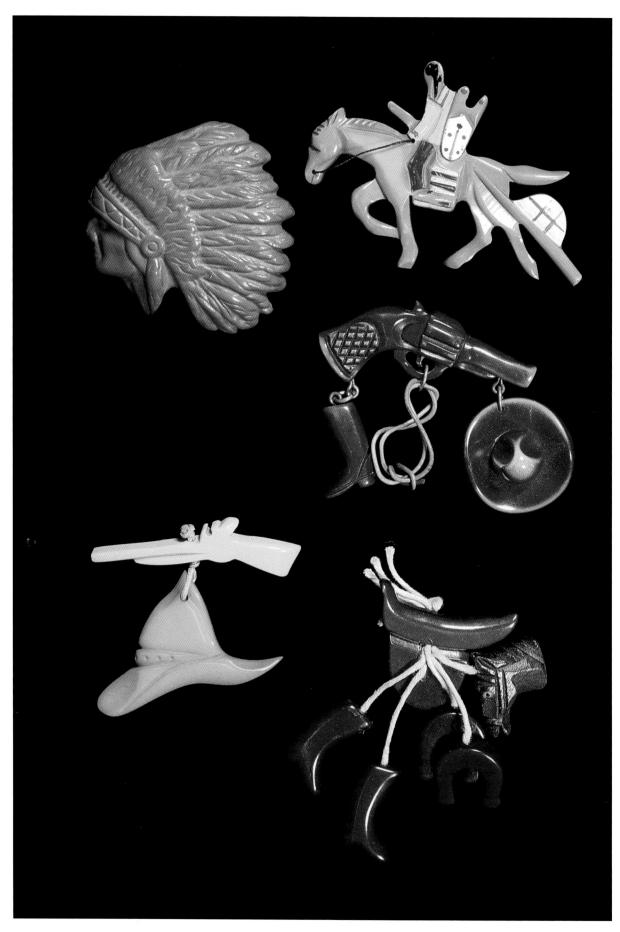

Wild West pins, c. 1930, in Bakelite, except for plastic saddle. $295-500.
Indian chief scarf slide, c. 1950s. $65

Novelty pins, c. 1935-42.
   Martha Sleeper fishing bag — good. $400
   ukulele — good. $375
   Martha Sleeper school pin — better. $750

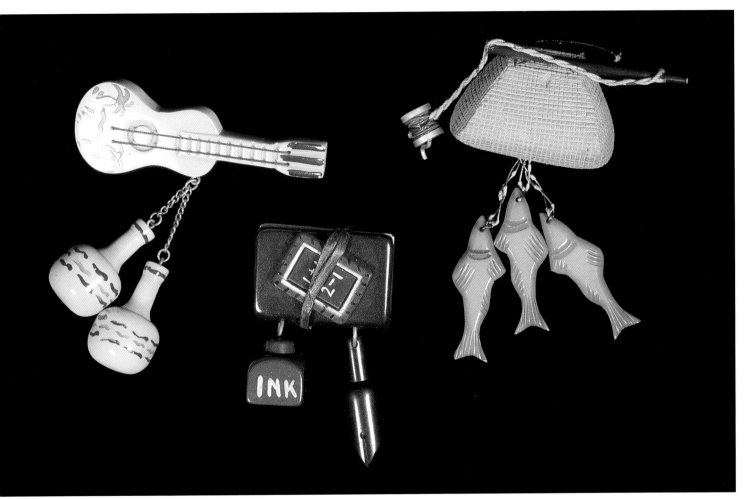

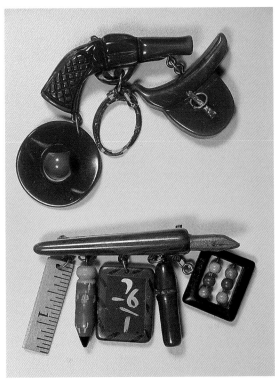

Pins, c. 1938. *Courtesy Judy Fitzpatrick.*
   pistol and charm — good. $595
   Martha Sleeper school — better. $1,200

Rider on burro pin, c. 1935.
Courtesy Judy Fitzpatrick. $300

Mexican figure pins, c. 1930s. $400 each, jointed $800

Mexican theme pins, c. 1935-40.
    laughing donkey — good. $300
    vaquero (cowboy) — better. $350
    sleeping man and palm — best. *Courtesy Beverly Ratner.* $425

Ethnic mask pins. African subjects were popular in the late 1930s.
  Bottom left: Lucite and Bakelite — good. $95
  Bottom right: Bakelite and chrome wire — better. $115
  Top: wood and Bakelite beads — better. $115

Bakelite pin with brass studs (missing brass ear and nose rings), c. 1928. $350

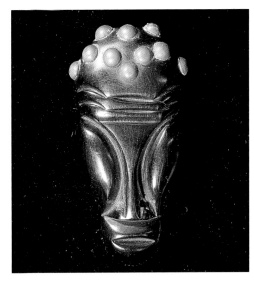

Hat pin, c. 1930. $550

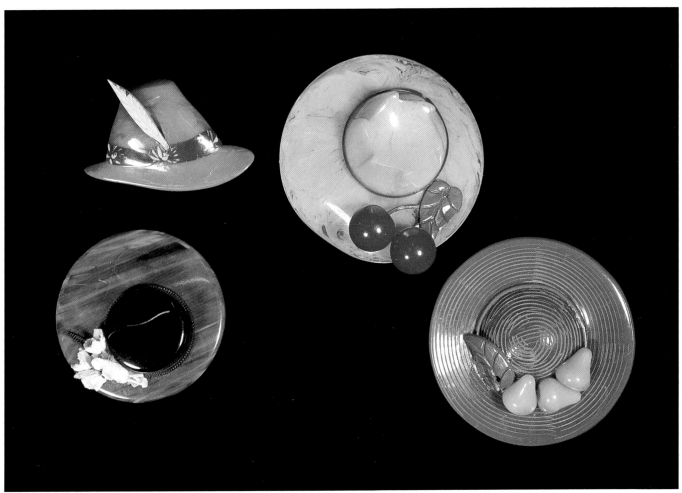

Hat pins c. 1930-37.
  cloth flowers — good: $175; ivory feather — better: $275; cherries — almost best: $600; pears — best: $750

Hat motif pins, c. 1935. *Courtesy Judy Fitzpatrick.*
    sombrero — good. $400
    painted polka dot — better. $550

Hand pins, c. 1937
    plain black hand — good. $300
    painted cream hand — better. $550

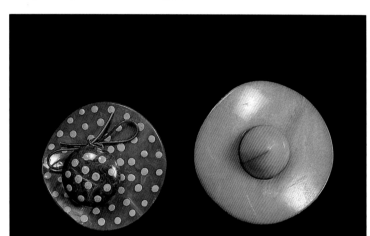

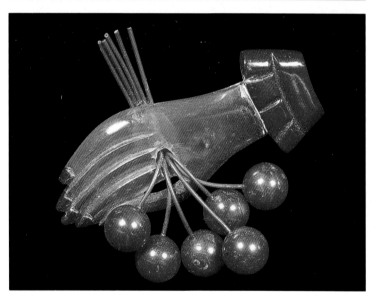

Plastic hand with Bakelite berries pin, c. 1939. $95

Red bow with berries pin, c. 1935. $400

Right:
Pin made to resemble ivory, c. 1940. $175

Below Left:
Hobé pin of enamel on brass with ivory spinner heads and Bakelite pieces under the heads. $150

Below Right:
Reverse.

Wood tulips on Bakelite pin, c. 1935. $150

Red Bakelite poinsettia on wood pin. $165

Bakelite on wood pin. $150

Apple juice leaf pin with reverse carving, c. 1930. $125

Leaves in a variety of colors, c. 1935. $85 each

Flower pin, c. 1935. $200

Flower pin, c. 1935. $185

Good, better, best pins, c. 1935, based on amount of carving.
Left to right: $150, $175, $200

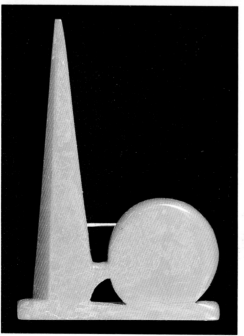

1939 New York World's Fair souvenir pin. The Fair included a Bakelite Pavilion, where tiny orange and blue Trylon and Perisphere pins embossed "Bakelite" were given away. The exhibit tied in plastic jewelry, notions, crib toys, and buttons with the concept of the future. $350

Christmas angel, c. 1939. It is unusual to find Christmas ornaments. $975

Opposite Page Top:
Novelty pins, c. 1940.
    clock with brass — good. $150
    Bakelite purse with plastic gloves — good. $150
    Bakelite match with plastic matchbook — better. $195

Pin of celluloid plastic with Bakelite root beer acorns and gold detailing, c. 1930 — good. $250

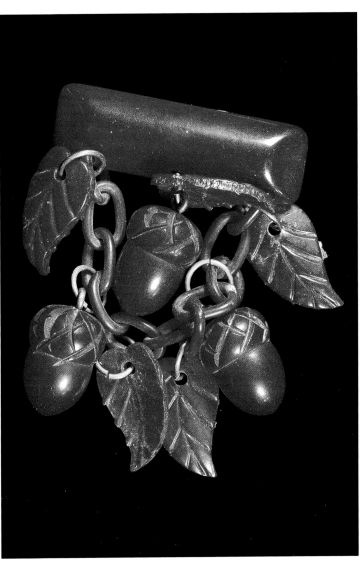

Red acorns with celluloid leaves pin, c. 1935 — better. $300

Resin wash acorns pin, c. 1935 — best. $675

Multi-vegetable pin, c. 1935-42. $425

Vegetable jewelry was popular during the war and sent the message to "eat your daily vitamins." These pins are c. 1935-42. *Courtesy Judy Fitzpatrick.*
    mixed vegetables — very good. $450
    single carrot — better. $600

Bakelite radishes on wood pin. $200

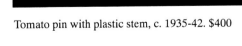

Tomato pin with plastic stem, c. 1935-42. $400

Top: green cherries on resin washed log pin.
Bottom: bar with eight red cherries pin.
Courtesy Judy Fitzpatrick. $350 each

NextPage:
Fruit pins, c. 1930s. *Courtesy Judy Fitzpatrick.*
oranges from log — good. $300
multi-fruit — better. $450
peaches from branch — better still. $750
strawberries from branch — best. $850
apples from branch — best. $950

Cherry pins, c. 1935-40.
carved with red log — good. $150
multi-fruit carved on red bar — better. $250
carved on green bar with laminated buttons — best. $325

Fruit pins and a clip, catalog 1935-40.
    pear clip — good. $250
    pear pin — better. $400
    apple pin — good. $200
    lemon pin — better. $375

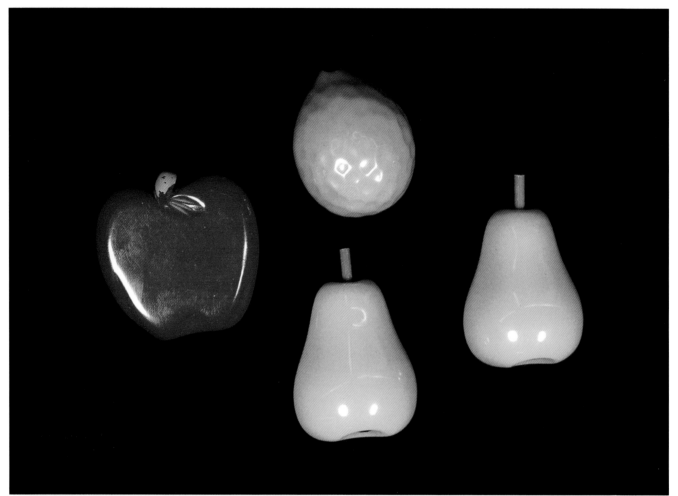

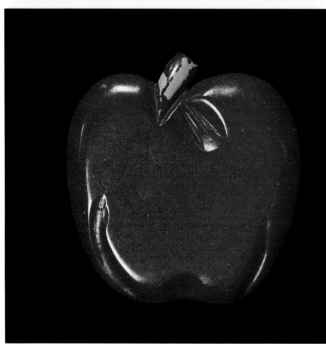

Apple pin, c. 1940. *Courtesy Judy Fitzpatrick.* $250

Root beer Bakelite pears on wooden leaves pin (you need to really be focused to get that one). $200

Raspberries on bar pin, c. 1935. *Courtesy Beverly Ratner.* $650

Pear cluster pin, c. 1930.
Courtesy Beverly Ratner.
$525

Currant clusters, c. 1935.
left, clip — good. $175
right, pin — better. $300

Small drum major hat pin, c. 1940.
Courtesy Judy Fitzpatrick. $175

Pin with painted dots, c. 1938. $225

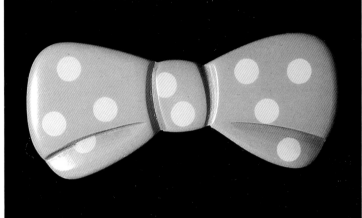

Carved pins. $50-150

Detail of $150 pin, c. 1930.

Bakelite pin, c. 1930. $200

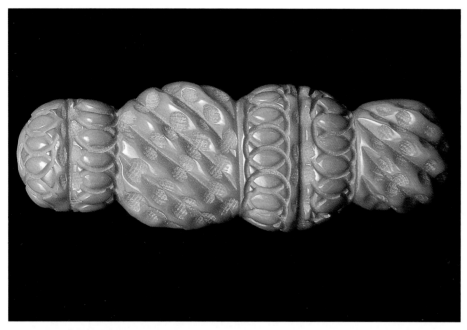

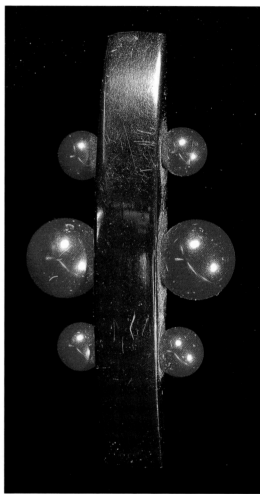

Two-color black and red pin, c. 1930. $250

Carved orange pin with seven Bakelite beads, c. 1935. *Courtesy Sheila Wolf.* $550

Philadelphia" pin, c. 1935. *Courtesy Judy Fitzpatrick.* $1,100

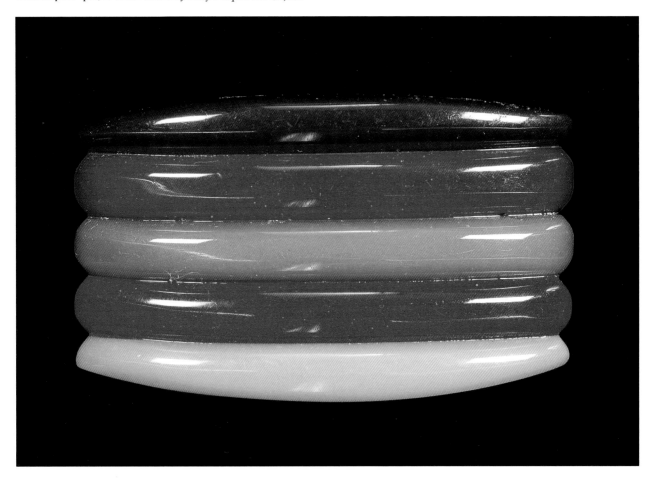

Reverse carved and painted pin laminated with Bakelite, c. 1935. $350

Apple juice reverse carved pins. *Courtesy Judy Fitzpatrick.*
   apple juice — good. $350
   opaque Bakelite with apple juice — better. $450

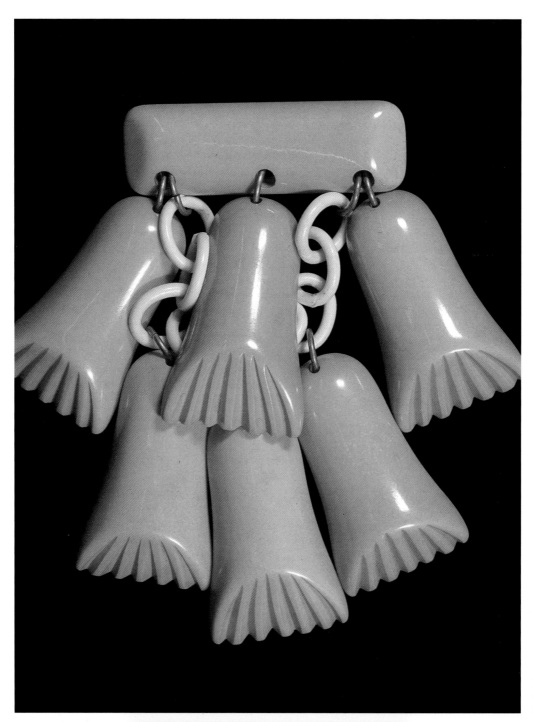

Pin with celluloid chain and Bakelite "bells," c. 1939. $425

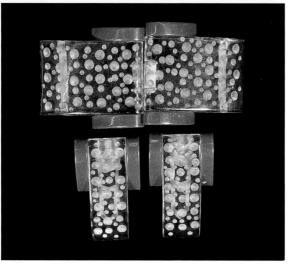

Dress buckle and matching clips of reverse carved apple juice Bakelite, c. 1938. $150 set

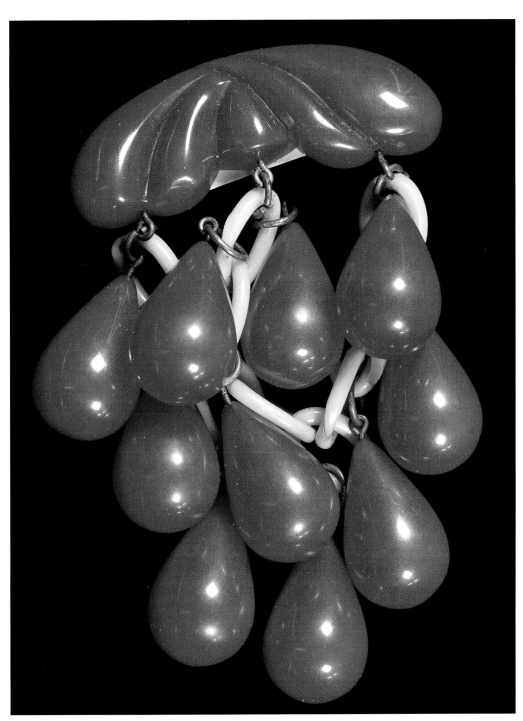

Orange pin with celluloid chain, c. 1935. $295

Bakelite and brass pin, c. 1930.
Courtesy Judy Fitzpatrick. $300

Necklaces, Clips, Etc.

Opposite Page:
 Fruit necklaces, c. 1935.
　　strawberries — very good. $750
　　mixed fruit — even better. $950 *Courtesy Beverly Ratner.*
　　mixed fruit with stems — best. $1,400

Above: Fruit necklaces, c. 1935.
　　red carved cherries — good. $450
　　oranges — better. $625
　　black carved cherries — better. $650

147

Cherry necklace with Bakelite leaves on a brass chain, c. 1935. *Courtesy Studio Moderne.* $200

Three-strand necklace resembling jade, c. 1930-35. $175

Charm necklace, c. 1935. Yellow hearts are resin wash over white Bakelite; the green heart is resin wash over butterscotch; jugs and beads are green, butterscotch, black, and red; the sombrero is laminated butterscotch and red. $700

Reverse carved apple juice pendant on celluloid chain. *Courtesy Studio Moderne.* $165

Necklaces with celluloid chains. Left: all celluloid, as Bakelite cannot be twisted. Right: Bakelite discs. *Courtesy Studio Moderne.* Left: $35; right: $125

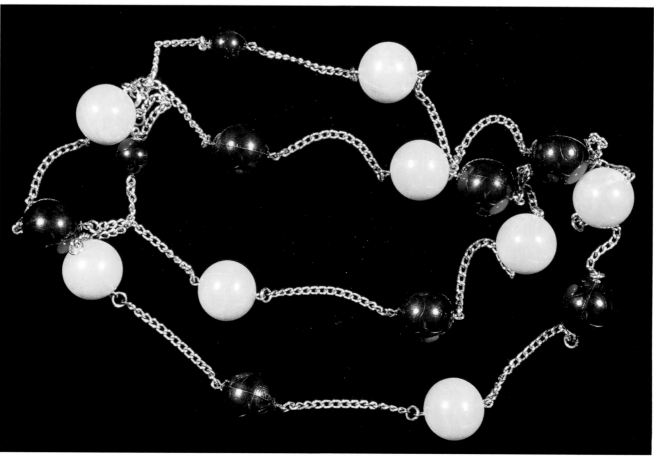

Below: Plastic and brass necklace of cream corn and black beads colored to look like Bakelite. *Courtesy Studio Moderne.* $150

Left: age-darkened teal bead necklace of older beads, but recently strung. $75
Right: cream color bead necklace, c. 1930s, made to imitate ivory. $175

Humpty Dumpty clip, c. 1940. $175

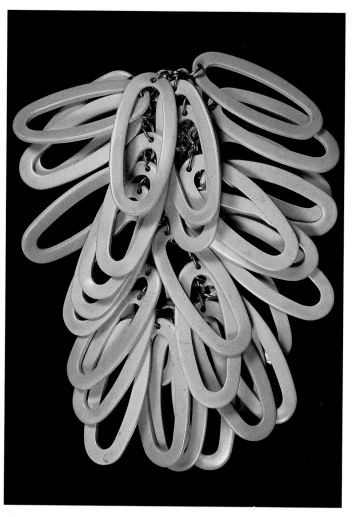

Thirty ovals on a clip, c. 1937. *Courtesy Beverly Ratner.* $300

Top, belt buckle; center, pair of clips; bottom, single clip. $55, $50, $125 respectively.

Clips. *Courtesy Shirley Friedland.*
 center — good. $25 right — better. $35 left — best. $100

Left to right: grape cluster clip, bar pin, and clip. *Courtesy Studio Moderne.* $175, $125, $35 respectively

Art Deco pins, c. 1930. *Courtesy Studio Moderne.*
left, hat pin with rhinestones — good. $55   black pin with rhinestones — best. $95
green clip with rhinestones — good. $35   brass pin — better. $85

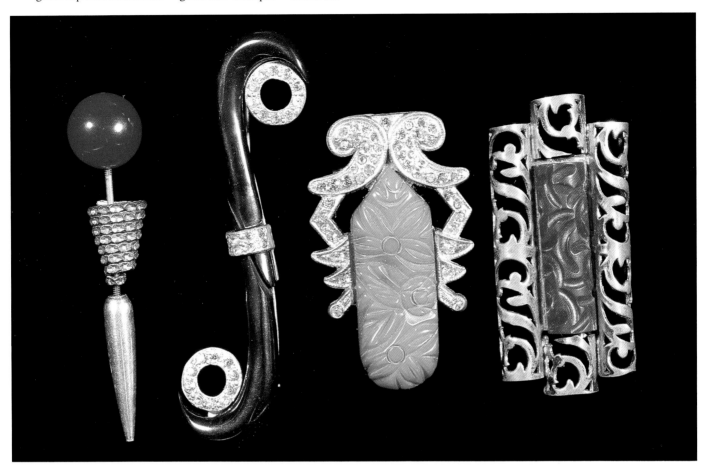

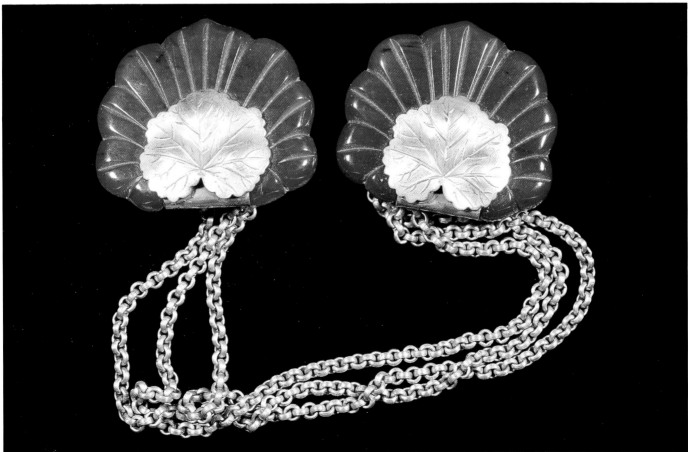

Sweater clip, c. 1940. *Courtesy Studio Moderne.* $125

Little lamb button. *Courtesy Judy Fitzpatrick.* $200

Assorted rings. *Bottom courtesy Shaboom's.* $100-200 each

Detail of the back of the lamb button.

Earrings to match 6-dot bracelets, c. 1935. $100

Opposite Page Upper Left:
Top left: burgundy bead and band bracelet which can be worn two ways. Top right: burgundy three cherries pin. Bottom: bar with eight cherries pin. *Courtesy Judy Fitzpatrick.* $195, $300, $600

Opposite Page Upper Right:
Sports pin $650; matching bracelet $295 (charm bracelets should have at least five charms).

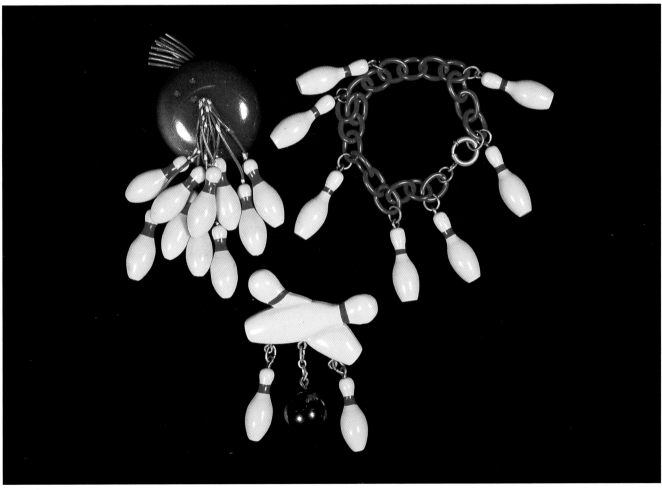

Two bowling pins and matching bracelet, c. 1938. $550 each pin, $250 bracelet

Left: poinsettia pin on wood. Right: red and green carved bangles. Pin: $165; bangles: $100 each

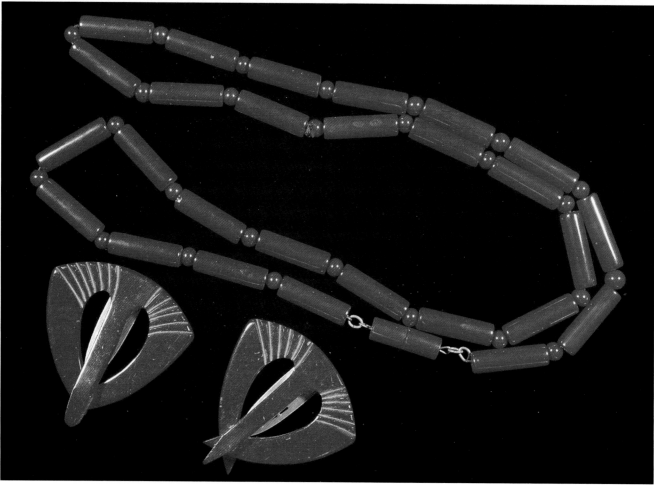

Bakelite necklace with Bakelite clasp, c. 1930; pair of clips, c. 1937.
Courtesy Studio Moderne. Necklace: $90; clips, $65

Dress made of Bakelite print fabric. *Courtesy Judy Fitzpatrick.*

Next page: Detail.

Objects

Although jewelry is probably the most collectible category of Bakelite items today, it only represented about ten percent of the Bakelite production of the late 1930s. Almost half of the Bakelite produced in 1937 was used to make buttons.

The first commercial use of Bakelite was for telephones in the United States and, then, in Europe. After the repeal of prohibition, cocktail shakers became status symbols and also added a touch of modern design to any decor — especially those of chromium with Bakelite. In the early 1930s, Schick's first electric shaver had a Bakelite casing and sold for a whopping $25, and Kodak introduced Walter Dorwin Teague's design, the "Baby Brownie" box camera, which sold four million at one dollar each. The Bakelite box camera was discontinued in the mid-1950s. The popular, often streamlined, Bakelite and Catalin radio cases were also phased out by the 1950s, when brightly-colored brittle plastics were used for the then fashionable angular and boxy designs.

The fashion for hot colors in the home was fueled by Bauer Pottery and Homer Laughlin's Fiesta ware. This helped to create a market for items like cutlery with catalin handles in coordinated colors. Corn holders became popular in the 1940s with the practice of backyard barbecuing, and the early corn holders were of Bakelite.

Bakelite could also be used for large items. Coffins were made in the 1930s, but understandably, were never produced on a large scale. The Queen Mary and Queen Elizabeth luxury liners had durable wipe-clean surfaces, such as wall paneling, flush doors, and bar counters, of a resin called Warite, which was made by the Bakelite Co. Perhaps the largest of the Bakelite items were the dance floors on which Fred Astaire reportedly liked to dance. The examples that follow, however, are some of the more popular small items, such as containers and games.

Unusual carved statuette of Ava Gardner from the film *A Touch of Venus*, c. 1948. *Identification courtesy of Bruce McClung.* 12-1/4 inch height. $425

Left:
Herpicide piece from barber shop of glass and Bakelite, embossed with "Herpicide" in Art Deco design, c. late 1920s. $75

Center:
Advertising pieces.
    Gypsy wine — good. $50
    Rupert's Ale, plastic
    and Bakelite — better. $120
    ashtray with "Bakelite" — best. $140

Top hat box, c. 1930, may have originally held perfume. $150

Opened box.

Box with Lucite bottom, possibly from a hobby kit, c. 1940, 3-3/4 inch diameter. $400

Left Center:
Tan mottled box with screw top, British, c. 1928, 4-3/4 inch diameter. $100

Right Center:
Boxes, c. 1930. *Courtesy Studio Moderne.* Large: $300; small: $150

Hobby kit box with decoupage top, c. 1940. $100

163

Biscuit or bon-bon box, English, c. late 1920s-30s, 10-1/8 x 7-1/4 x 4 inches. $2,000

Top of box.

Detail.

Boxes, c. 1930s.
    back hinged box — good. $50
    small jewelry presentation box, back hinged — better $100
    "Cleopatra" box by General Electric, made of Textolite — best. $400

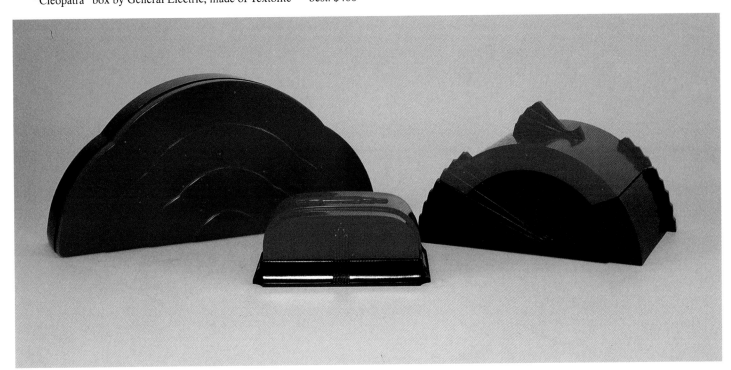

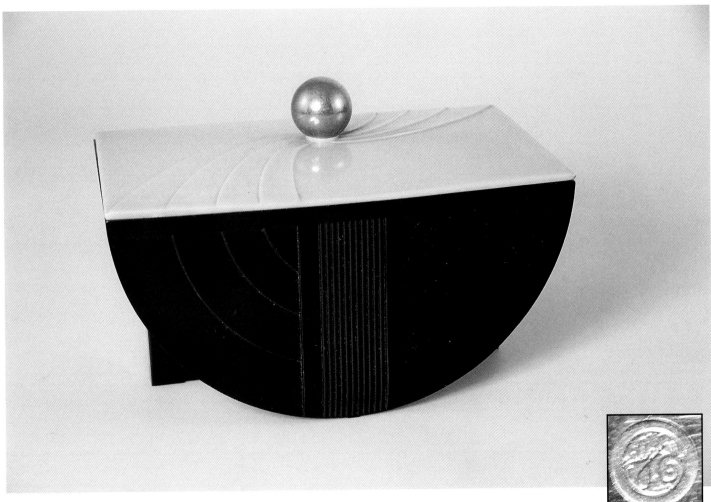

Textolite (G.E. trademark for phenolic resin) and plastic top box made for Chicago Century of Progress World's Fair in 1933, stamped "G.E." and one of two styles made. $165

Bottom of box.

Travel container and holder for darner, English. $125

Contents of travel container: brush and tooth brush, English, c. late 1920s. $150

Seth Thomas clock with Bakelite face, c. 1930.
Courtesy Beverly Ratner. $300

Left: butterscotch Bakelite portable ashtray that closes into a box. Center: Bakelite vase. Right: unusual mouse lighter made of palm nut and Bakelite, marked "Austria." This is one of many items such as lighters, match strikers, and toothpick holders that were made of these materials. Left to right: $75, $85, $120

"Novelview" with original box. $22

Bakelite binoculars with leather case. $100.

Left: Bakelite pen holder with pen. Center: chrome and Bakelite picture frame.
Right: cigarette or card box, shaped like a book, of base metal with Bakelite and wood liner.
Left to right: $110, $125, $100

Book rack with World War II scenes of men and planes on front and side. $160

Detail.

General Electric tea kettle with Bakelite handle in streamlined design of late 1930s. $85

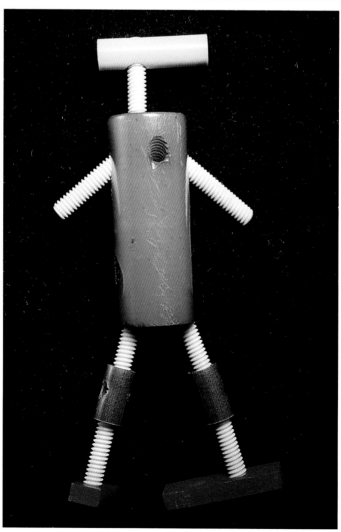

Early toy, similar to "Tinker Toys," of Bakelite with plastic connectors, c. 1930. $100

Brandy or whiskey glass holder, c. 1935. $185

Bakelite billiard balls, c. 1930. $100

Assortment of buttons. Figural buttons are often referred to as "realistics."

Buttons on original cards. By 1937, of the 5-1/2 million pounds of cast phenolic resin produced, almost half was for button manufacture and only 10% for jewelry.

Napkin rings.
    thick rabbit — good. $85
    sitting airedale — better. $115
    Mickey Mouse — best. $165

Figural napkin rings, c. 1940. The rocking horse is the least common. $60-175

Salt and pepper sets, c. 1935.
    pair with tray — good. $95
    mushrooms — better. $100
    cherries with chrome stems — best. $150

Figural pencil sharpeners, c. 1930-42. Small items with perfect decals have become very collectible, especially a large variety of Disney characters. $35-150

Selected Bibliography

Books

Becker, Vivienne. *Rough Diamonds: The Butler & Wilson Collection.* New York: Rizzoli, 1990.

Davidov, Corrine and Ginny Redington Dawes. *The Bakelite Jewelry Book.* New York: Abbeville, 1988.

DiNoto, Andrea. *Art Plastic: Designed for Living.* New York: Abbeville, 1984.

Fink, Nancy and Maryalice Ditzler. *Buttons.* Philadelphia: Running Press, 1993.

Grasso, Tony. *Bakelite Jewelry: A Collector's Guide.* Seacaucus, New Jersey: Chartwell, 1996.

Katz, Sylvia. *Plastics: Common Objects, Classic Designs.* New York: Harry N. Abrams, 1984.

Kelley, Lyngerda and Nancy Schiffer. *Plastic Jewelry* . Atglen, Pennsylvania: Schiffer, 1996.

McNulty, Lyndi Stewart. *Price Guide to Plastic Collectibles.* Radnor, Pennsylvania: Wallace-Homestead, 1992.

Miller, Harrice Simons. *Costume Jewelry.* New York: Avon, 1994.

Mumford, John Kimberly. *The Story of Bakelite.* New York: Robt. L. Stillson Co., 1924.

Shields, Jody. *All That Glitters: The Glory of Costume Jewelry.* New York: Rizzoli, 1987.

Sparke, Penny, ed. *The Plastics Age: From Modernity to Post-Modernity.* London: Victoria & Albert Museum, 1990.

Articles

"Beyond Bakelite." *Women's Wear Daily.* Dec. 7, 1990.

DiNoto, Andrea. "Bakelite Envy." *Connoisseur.* July 1985.

Fadem, Lloyd and Stephen Fadem, M.D. "Recrafted Bakelite." *Echoes Report.* Vol. 5, No. 1, Summer 1996.